The Two-Income Trap Escape Plan

Managing Money as a Household

Joseph Trek

Published by NobleTrex Press

For permissions and other inquiries, write to:

P.O. Box 3132, Framingham, MA 01701, USA

Contents

Introduction

Here is a paradox that keeps me up at night, and I suspect it keeps you up, too.

Imagine a household in 1975. A single earner works at a factory or an office. The other spouse stays home. They own a modest three-bedroom house, drive one car, and send two kids to the local public school. They save a little bit every month. They aren't rich, but they sleep well.

Now, fast forward to today. You and your partner both work. You are educated, ambitious, and likely earning a combined salary that, even adjusted for inflation, dwarfs what that 1975 household brought in. You have two cars, smartphones that contain the sum of human knowledge, and a house that is statistically 1,000 square feet larger than the one from fifty years ago.

On paper, you are a roaring success. You are the embodiment of the economic dream.

So why does it feel like you are one bad Tuesday away from total collapse?

Why does a broken transmission, a surprise medical bill, or a rumor of layoffs at the office send a cold spike of adrenaline through your chest? Why, despite working twice as hard and earning twice as much, do you feel half as secure?

Welcome to the Trap.

It isn't a failure of character, and it certainly isn't because you buy too many lattes or avocado toast. That is the lazy narrative of cable news pundits. The reality is much colder and purely mathematical. You feel fragile because you *are* fragile.

Over the last few decades, we silently shifted the goalposts of household finance. We moved from a model of *resilience*-where one income paid the bills and the second (if it existed) was pure surplus-to a model of *maximization*. We took that second income and immediately committed it. We bought the house that required two salaries to qualify for. We bought the two cars necessary to commute to those two jobs. We signed up for the childcare required to free up the time to work those jobs.

We built a financial machine that generates massive horsepower but has no brakes and no suspension. It runs beautifully on a smooth, flat road at 100 miles per hour. But the moment you hit a pothole-a recession, a sick parent, a burnout diagnosis-the wheels come off.

This book is not about how to get rich quick. It is not about picking the next hot stock or crypto-token that will let you retire on a yacht next Tuesday. If you are looking for that, put this back on the shelf; I cannot help you.

This book is about engineering. It is about taking the high-performance, high-risk machine you have built and retrofitting it into a tank. It is about escaping the trap where you are forced to work to pay for the things that allow you to work. It is about staring at the ceiling at 3:00 AM and realizing, for the first time in years, that if you lost your job tomorrow, your family would be okay.

The Myth of the "Number"

For years, the financial industry has sold us a lie called "The Number." You've seen the commercials. A grey-haired couple walks along a vineyard, smiling at a sunset, and a gentle voiceover asks, "What is *your* number?" The implication is that financial success is a giant pile of cash sitting in an account somewhere, waiting for you to turn sixty-five.

The problem with focusing on The Number is that it ignores the thirty or forty years of life you have to survive to get there. It treats your life as a waiting room for retirement.

Real financial success isn't a pile of money; it's a flow. It's the relationship between what flows in and what *must* flow out.

In these pages, we are going to obsess over a different metric. I call it the "Minimum Monthly Nut." This is the absolute minimum amount of money your household needs to keep the lights on and the roof up. Most dual-income households today have a Nut that is terrifyingly close to their total take-home pay. They are running at 95% capacity. They have no slack.

When you have no slack, you have no choices. You cannot leave a toxic boss because the mortgage is due on the first. You cannot take a month off to care for a sick child because the car payments don't stop. You cannot start a business because you need the health insurance. You are wealthy in assets but impoverished in options.

The escape plan we are going to build together focuses on widening the gap between your income and your Nut. But-and this is critical-we aren't just going to do it by shouting at you to stop spending money. Willpower is a finite resource;

eventually, it runs out. Instead, we are going to use system design. We are going to restructure your household balance sheet so that saving isn't a chore you have to remember to do, but an automatic byproduct of how you live.

The Household as a Corporation

To escape the trap, we have to stop treating household money like a chaotic emotional experiment and start treating it like what it actually is: a small business.

Your household is an organization. It has revenue (salaries), overhead (mortgage, food), assets (home, investments), and liabilities (debt). It has shareholders (you, your partner, your kids). It faces risks (inflation, lawsuits, health).

Yet, most couples run this organization with a management strategy that boils down to "checking the bank app in the parking lot to see if the card will decline."

It's time to professionalize.

We are going to walk through the creation of a "Household Operating System." This sounds unromantic, I know. But there is nothing romantic about arguing over credit card bills in the kitchen while the pasta boils over. There is nothing romantic about hiding purchases from your spouse. True romance is the peace of mind that comes from transparency.

We will look at how to structure your accounts-not just "joint vs. separate," but an architectural strategy that allows for autonomy without secrecy. We will discuss the "Money Meeting," a ritual that usually inspires dread, and transform it into a twenty-minute sync that actually reduces your cortisol levels.

We are going to look at the data, stripped of judgment, and make decisions based on what the math says, not what your neighbors are doing.

The Silent Invisible War

While we are fixing your internal systems, we also have to acknowledge the external war being waged against your wallet. You are fighting an asymmetric battle.

On one side, there is you: tired, busy, perhaps balancing a laptop on one knee and a toddler on the other. On the other side, there are billion-dollar corporations employing Ph.D. psychologists and algorithmic data scientists with one singular goal: to separate you from your money.

They have weaponized convenience. They have turned spending into a frictionless slide. One-click ordering, subscription models that bleed you ten dollars at a time, "Buy Now, Pay Later" schemes that maximize your dopamine today and outsource the pain to your future self.

And then there is the silent thief: Inflation. We usually think of inflation as the price of milk going up. But the more dangerous form is "lifestyle inflation." It's the creeping normality of luxury. It's the moment when a treat becomes a habit, and a habit becomes a necessity. It's the reason why a promotion at work rarely leads to more savings-it just leads to a nicer car and a more expensive brand of gin.

We are going to learn how to identify these forces and build defenses against them. You cannot rely on self-control to fight an algorithm. You need better barriers. We will talk about "friction"-deliberately making it harder to spend money so that your rational brain has a chance to catch up with your impulsive brain.

The Fixed-Cost Buzzsaw

The centerpiece of our strategy-and the part that will likely make you the most uncomfortable-is the section on fixed costs.

If you read most personal finance books, they obsess over the variable costs: the lattes, the streaming services, the gym memberships. And sure, those matter. But they aren't what kill you.

What kills you are the decisions you sign your name to. The thirty-year mortgage. The five-year car loan. The student debt.

We are going to have a very honest conversation about housing. For three generations, we have been told that a house is the best investment you can make. We have been told that "renting is throwing money away." I am going to show you the math that proves why this is often dead wrong, and how the obsession with homeownership has trapped millions of families in illiquid, expensive assets that prevent them from building real wealth.

We will look at cars, the great destroyers of middle-class wealth. We will look at the student loan crisis not as a political issue, but as a balance sheet liability that needs a ruthless elimination strategy.

This is the "Escape" part of the Escape Plan. It involves making hard choices to lower that Minimum Monthly Nut. It might mean driving a car that doesn't impress the valet. It might mean living in a neighborhood that is "up and coming" rather than "already arrived."

But the prize? The prize is a life where you don't need to check your bank balance before buying groceries. The prize is the ability to walk away from a job that makes you miserable.

Investing: The Engine, Not the Casino

Once we have stopped the bleeding and stabilized the patient, we can finally talk about growth.

Investing is usually presented as a dark art, accessible only to men in fleece vests watching six computer monitors. In reality, successful investing is incredibly boring. It is about understanding a few powerful forces-compound interest, diversification, and cost control-and then doing as little as possible.

We will strip away the jargon. I won't tell you how to beat the market, because you can't. Instead, I will show you how to capture the market's returns without paying a fortune in fees. We will discuss the "Investor's Toolkit"-the index funds, the tax-advantaged accounts, and the automated flows that turn money into more money while you sleep.

But more importantly, we will talk about the psychology of the investor. The biggest threat to your portfolio is not the Federal Reserve or the price of oil; it is the person looking back at you in the mirror. We are hardwired to buy high when we feel greedy and sell low when we feel scared. We are going to build a "Family Investment Policy"-a constitution for your money-that ties your hands so you don't do something stupid when the news gets scary.

Why This Matters Now

You might be thinking, "This sounds like a lot of work. Can't I just keep winging it?"

You can. But the margin for error is shrinking.

We live in an era of disappearing guarantees. Pensions are a relic of history. Social Security is under pressure. The cost of healthcare and education is rising faster than wages. The corporate ladder has been replaced by the gig economy.

No one is coming to save you. The cavalry is not over the hill. You are the cavalry.

This realization can be terrifying, or it can be empowering. If no one is coming to save you, then you have the ultimate permission to build your life exactly the way you want it. You don't have to follow the standard script-college, debt, marriage, mortgage, bigger mortgage, retirement, death-because that script is broken.

When you manage your household money with clarity and purpose, you aren't just changing numbers on a spreadsheet. You are buying time. You are buying the ability to be present with your children without worrying about the electric bill. You are buying the bandwidth to be a better partner, a better friend, and a better citizen.

Money, at its core, is stored energy. Right now, your energy is leaking out through cracks in your foundation, fueling a lifestyle that doesn't actually make you happy. We are going to seal those cracks. We are going to harness that energy.

We are going to take the anxiety that sits in the pit of your stomach and replace it with a plan.

The journey we are about to take is not about deprivation. It is not about living in a cold house and eating beans (unless you really like beans). It is about *alignment*. It is about ensuring that every dollar you earn is working toward the life you actually value, rather than the life you think you're supposed to have.

The trap is open. The exit is right here. Let's get to work.

Chapter 1

The New Rules of Household Money

Most of us play the money game without ever reading the rulebook. We chase higher salaries only to feel more squeezed, wondering where it all goes. This chapter hits pause on the rat race to explain the invisible forces-inflation, time, and compounding-that govern your wealth. We'll replace the outdated goal of 'keeping up' with the radical power of 'defining enough,' giving you the physics you need to build a household that doesn't just survive, but thrives.

1.1 Defining Success: Goals and Values

On paper, many households look "fine." Two paychecks. A home. Two cars. A vacation photo once a year. Maybe even a retirement account that gets a contribution when things are calm. And yet the lived experience is tight-chested: the sense that one surprise could knock the whole setup sideways. The strange part is that the stress often *increases* right alongside the outward signs of success. More income, more obligations. More comfort, less room to breathe.

That contradiction usually isn't a math problem. It's a definition problem.

Most of us inherit a default definition of financial success without ever agreeing to it. It's ambient. It leaks in through group chats, family expectations, neighborhood norms, and the quiet comparison engine that runs in your head while you're waiting in line for coffee. The default definition tends to be public and visible: the size of your house, the newness of your car, the brand names you wear, the "normal" activities your kids do, the restaurants you can say yes to without checking your bank app.

Public success has one major advantage: it's easy to measure. You can point to it.

It also has one major downside: it doesn't necessarily buy the thing you actually want.

A better definition of financial success starts in a less glamorous place: *What do you want your money to do for you?* Not what you want it to say about you. Not what it should look like from the outside. What it should *do*. The best answers tend to be quietly powerful: sleep, options, stability, time, generosity, dignity, the ability to say no, the ability to handle a crisis without borrowing, the ability to help your future self without resenting your present.

When a household rewrites success in those terms, money becomes less of a scoreboard and more of a tool. That shift is the difference between "We earn a lot" and "We're okay even if one of us can't work for a while." Between "We upgraded" and "We can make a decision without panic." Between "We keep up" and "We can change direction."

The Most Expensive Words in Personal Finance: ``We Deserve It"

There's a moment most couples recognize, even if they don't talk about it. It's the moment when something stops being a choice and turns into a baseline.

At first, it's a treat: the meal delivery service during a brutal work month, the bigger apartment because the commute was wrecking your life, the new SUV because you were tired of feeling cramped. The reasoning is understandable. You *do* work hard. Life *is* stressful. Money is supposed to help. And then the baseline sets like concrete. Soon, you're not paying for convenience—you're paying to avoid discomfort you can't imagine returning to.

That is lifestyle inflation, and it usually sneaks in wearing the mask of "we're just being practical."

But the deeper issue is that "practical" often means "socially normal." If everyone in your peer group seems to have the bigger house and the travel sports and the upgraded kitchen and the summer camp, opting out can feel like choosing deprivation for no reason. The feeling is powerful because it's not really about granite countertops. It's about belonging.

Here's the catch: a household can belong to a social world whose "normal" requires financial fragility.

And financial fragility is expensive. Not always in dollars, but in the way it corners your choices. It can trap you in jobs you've outgrown, keep you from relocating, turn minor health issues into crises, and make every unexpected expense feel personal—as if you failed a character test—instead of what it really is: life happening.

Redefining success is how you stop buying fragility.

Two Definitions of ``Rich" (Only One Matters on a Tuesday Night)

There are two kinds of rich.

The first is the kind you can photograph. Big house. Nice car. A kitchen that looks like it belongs in a magazine. It's real wealth sometimes—but not always. Plenty of households look rich and live one paycheck away from chaos. The bills still come. The childcare still costs what it costs. The debt still needs to be fed every month. The stress is quieter than a repo truck, but it's there.

The second kind of rich is harder to see: a household that can absorb a shock without borrowing, can take a risk without risking everything, can handle a job loss without immediate panic, can say no to a demanding boss, can choose a better-fit school or a different neighborhood or a less prestigious title because they're not financially cornered.

That kind of rich is what most people are actually chasing, whether they know it or not. It's a form of safety and choice. It makes ordinary life feel lighter. It's noticeable on a random Tuesday night when the car makes a weird noise, or the pediatrician says "we need a specialist," or your manager hints at layoffs. If your stomach drops, you're not lacking intelligence; you're lacking slack.

Slack is the unsung luxury. It's the space between your income and your commitments. It's margin in your calendar and margin in your budget. It's what turns money from an anxiety trigger into a stabilizer.

If you want a definition of success you can build a plan around, start with this: *Success is having enough slack to make good decisions under stress.*

Values: The Invisible Budget Categories

People often think of "values" as something vague and inspirational, like a poster in a hallway. But values are brutally practical. Your values determine what you protect when you can't protect everything. They determine what you say yes to when the calendar is full, and what you cut when money gets tight.

Here's the problem: most households have never said their values out loud in a way that connects directly to spending. They have a mental list of things that feel important—family, health, education, faith, freedom, security—but the budget tells a different story.

Budgets are honest in a way we rarely are.

If your bank statements show you spend heavily on convenience purchases that reduce daily friction, one of your values might be *ease* or *time* or *peace at home*. That isn't shallow. It's human. If you spend heavily on travel, you might value *novelty* or *connection* or *memories*. If you spend heavily on the appearance of success, you might value *status*—but it might also mean you value *belonging* and you've been taught that belonging must be purchased.

Values aren't the issue. Unexamined values are.

The goal isn't to force your household into a moralistic spending pattern. The goal is to make sure the dollars leaving your account are buying the life you actually want, rather than the life you feel obligated to display.

A household that knows its values can do something most households can't: cut spending without feeling punished. Because the cuts are not random. They're targeted. They're surgical. They remove costs that don't serve the household's chosen definition of success.

A Case Study: The ``Perfect" Life That Felt Like a Trap

Consider a couple I'll call Maya and Chris. Mid-30s, two professional incomes, two kids, suburban home. They were doing "everything right" by the default script. The house was in the right school district. The cars were reliable and not embarrassingly old. The kids had activities. Their friends were similar: same stage, same pace, same constant low-grade exhaustion.

Their financial numbers weren't catastrophic. They were contributing to retirement accounts, though not as much as they wanted. They paid their bills on time. They weren't drowning in credit card debt. From the outside, they were the kind of household people assume is stable.

But Maya described their finances with one word: *brittle*. If Chris lost his job, they'd last maybe two months before the stress became a crisis. If Maya took a lower-paying role with better hours, their whole system would wobble. They had built a life that required both incomes at full speed, indefinitely.

When we laid out their spending, the surprise wasn't the big bills. It was the number of medium bills: subscriptions, activities, upgraded phone plans, frequent takeout, the "nice" grocery store, the home services that felt essential because they were always behind. None of it was outrageous. In aggregate, it was suffocating.

So we tried something that sounded almost too simple: we wrote down their values in plain language and forced specificity.

They thought "family" was a value. That's everyone. The question is: *what does family look like in dollars and hours?* For them, family meant two things:

- eating dinner together most nights without rushing, and
- having the option for one parent to step back from work temporarily if a kid needed extra support.

Those are values you can build with money. They require slack, not upgrades.

They also listed what they realized they *didn't* value as much as they were paying for: being in the "right" neighborhood at any cost, keeping up with friends' activities calendar, and driving cars that felt like a personal brand statement.

That clarity changed everything, not because it revealed some secret hack, but because it gave them permission to stop chasing the wrong finish line. They downsized two major obligations over the next year: they moved to a slightly less coveted neighborhood (still safe, still good schools, just not the one that won cocktail-party approval), and they reduced the kids' activities to the ones the kids truly loved. They also implemented a "two nights" rule: two weeknights per week were protected from errands and commitments. That forced them to simplify.

The result wasn't deprivation. It was relief. Their monthly slack grew, their emergency fund stopped being a fantasy, and the household finally had the option Maya wanted: she moved to a less demanding role without the decision feeling like a cliff jump.

The point of that story isn't that everyone should move or cut activities. The point is that when success is defined socially, the spending will be social. When success is defined personally, the spending becomes strategic.

Opportunity Cost: The Phrase That Turns Spending Into a Choice Again

Opportunity cost is the adult version of the question your parents asked: "What else could you do with that money?"

In finance, it means this: every dollar can only be used once. When you spend it on one thing, you are not spending it on something else. That "something else" is the opportunity you gave up.

Opportunity cost sounds academic until you use it on a real purchase. Then it becomes a flashlight.

Suppose you're about to buy a 900 stroller, or a 1, 800 phone upgrade, or a 3, 500 family vacation, or a 650 monthly car payment instead of a 350 one. The default way to evaluate it is emotional and immediate: *Do we want it? Will it make life easier? Do we deserve it? Is it normal?*

Opportunity cost adds a second question, and it changes the outcome more often than people expect:

> What future problem would this money have solved if we didn't spend it this way?

That question is powerful because it makes the tradeoff visible. Money choices feel painful when they feel like denial. They feel empowering when they feel like a trade you chose.

Here's what opportunity cost does for a two-income household, specifically: it helps you stop using the second income to finance a more complicated life, and start using it to buy resilience. A second income can be a ladder out of fragility. But it can also become the fuel that inflates fixed costs—bigger mortgage, higher childcare expectations, car payments, subscriptions—until both incomes are required just to keep the machine running.

Opportunity cost is how you notice the machine.

A Concrete Example: The Car Payment That Ate a Safety Net

Imagine you're deciding between two car options:

- Car A: 350/month payment
- Car B: 650/month payment

The difference is 300/month.

If you only look at the cars, 300/month can seem like "the price of comfort" or "the price of safety" (whether or not it truly is). If you look at opportunity cost, 300/month becomes a lever.

Over three years, 300/month is 10, 800. Over five years, it's 18, 000.

Now translate that into household resilience: 18, 000 could be a meaningful emergency fund. It could cover a deductible and out-of-pocket medical max in a high-deductible year for many plans. It could keep you from carrying a credit card balance through a layoff. It could be a cushion that lets one person take unpaid leave when a family member needs care. It could be the difference between panic and options.

This doesn't mean Car B is "wrong." It means Car B is not just a car. It's also the decision to *not* buy those other outcomes.

Opportunity cost isn't a guilt trip; it's an honesty tool. It forces your household to answer: *What are we buying with this higher fixed cost, and is it worth the other life we're not buying?*

The Status Tax (And Why It's So Hard to Notice)

Status spending is uniquely tricky because it often masquerades as something else.

A bigger house can be framed as "for the kids." A luxury car can be framed as "reliable." A pricey neighborhood can be framed as "the schools." Sometimes those reasons are true. Sometimes they're partially true. But the status component matters because it changes your tolerance for cost. Status makes you stretch further than you otherwise would, because the purchase is not just functional—it's symbolic.

Symbols are expensive.

They're expensive not only because they cost money, but because they tend to become fixed costs. And fixed costs are the trapdoor in a two-income trap. Variable spending can be reduced in a bad month. Fixed costs don't care that you had a bad month.

When your definition of success is public, you pay a recurring status tax: the ongoing cost of staying in the visible version of "doing well." It's not just the initial purchase. It's the maintenance: higher insurance, higher taxes, higher expectations, the constant upgrades to keep the symbol from looking dated.

The antidote isn't to reject all nice things. The antidote is to buy nice things *on purpose*, with eyes open, in a way that doesn't turn your household into a high-speed treadmill.

A Simple Household Exercise: ``Enough" With Teeth

The word "enough" can sound like settling. But in practice, "enough" is a line you draw so you can stop negotiating with yourself every month.

Here's an exercise that works because it's specific:

1. Write down the *top three outcomes* you want your money to buy over the next 24 months. Not purchases—outcomes. Examples: "six months of emergency savings," "one parent can reduce hours," "pay off the last credit card," "save for a move," "more time at home," "less conflict about money."
2. For each outcome, write the *price tag* in dollars and the *price tag* in monthly cash flow. (Some goals are one-time; others require ongoing room in the budget.)

3. Now pick one spending category that is mostly about appearance or convenience and ask: *If we cut this in half, which outcome becomes real?*

This works because it ties cutting back to getting something you actually want. Most budgets fail because they feel like punishment plans. This turns budgeting into a trade: you're not "giving up restaurants." You're buying the ability to handle a crisis without debt. You're buying a calmer home. You're buying options.

And if you do decide to keep the spending, that's fine too—because you kept it with a clear-eyed understanding of what it costs you in the only currency that matters: your future choices.

The Takeaway: A One-Sentence Filter for Every ``Should We?"

For the next purchase that makes you hesitate—the upgrade, the trip, the remodel, the recurring service, the new monthly payment—use this filter before you talk yourself into it:

> If we spend this, what problem are we choosing *not* to solve—and are we comfortable with that trade?

Write your answer down like an adult signing a contract with your future self. If you still want the purchase after you name what it displaces, you're spending with values. If you don't, you just found money for the life you actually want.

1.2 The Language of Money: Time Value

A strange thing happens when money crosses time: it changes character.

A dollar sitting in your checking account today can pay a bill, prevent a late fee, cover a surprise copay, or keep you from putting groceries on a credit card. That same dollar promised to you *later* might be perfectly real—but it can't stop a problem that arrives *before* it does. The timing is the whole point. Money isn't just an amount; it's an amount *and a date*.

That idea has a plain name: *time value of money*. It simply means that a dollar today is worth more than a dollar tomorrow because today's dollar can be used (or invested, or used to avoid borrowing) in the meantime. Time gives money *options*. And options are what households run out of first when life gets complicated.

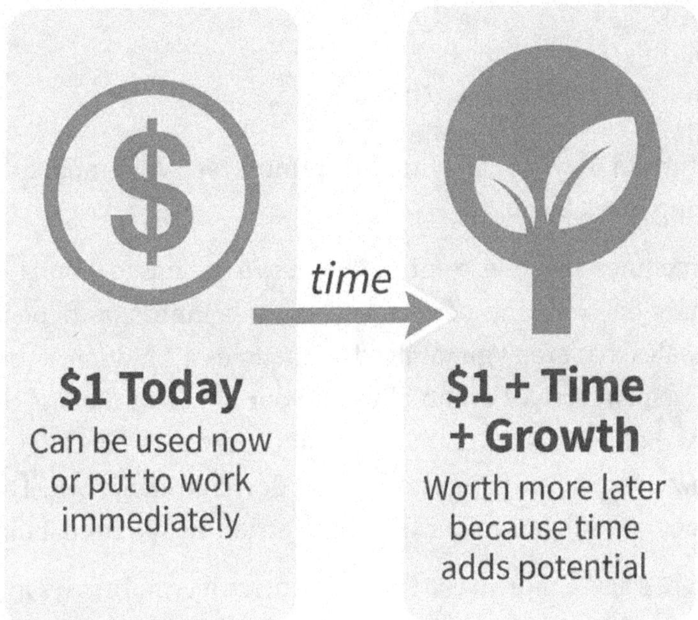

If that sounds like investing talk, it's broader than investing. Time value shows up in the most ordinary decisions a household makes: whether to keep cash on hand or send every spare dollar to debt, whether to take a signing bonus or a slightly higher salary, whether to prepay a year of insurance for a discount, whether to buy appliances on a "same as cash" plan, whether to choose a job with a higher 401(k) match but a longer commute, whether to refinance a mortgage, whether to lease a car, whether to accept a payment plan at the hospital.

Most financial stress comes from decisions made without translating money into a common language. Time value *is* that language. It lets you compare apples (money now) to oranges (money later) without guessing.

Why ``Later" Is Riskier Than It Looks

Future money comes with fine print, even when nobody is trying to trick you.

Sometimes the fine print is obvious: a promised bonus depends on company performance, or a commission depends on sales that aren't guaranteed. Sometimes it's hidden: a reimbursement arrives eventually, but your credit card bill is due now. Sometimes it's emotional: you plan to save "once things slow down," but things never slow down on their own. Time is not neutral. It's a filter that turns certainty into probability.

There's also a more basic truth: the future has more ways to go wrong than the present. Your car can break down. Your hours can get cut. A parent can get sick. Childcare can fall apart. You can be fine and then not fine. Money that exists today has one huge advantage over money that might exist later: it can respond to reality.

This is why liquidity (cash you can access quickly without penalties) has a value that isn't captured by spreadsheets. Liquidity is the household version of being able to breathe through your nose. You don't admire it when you have it. You panic when you don't.

Time value doesn't tell you to hoard cash forever. It tells you to respect what cash *does* for a household: it buys time, options, and the ability to avoid expensive forms of desperation.

Present Value: Translating a Future Promise Into Today's Dollars

When people say, "A dollar today is worth more than a dollar tomorrow," the natural question is: *How much more?*

That's where a simple tool comes in: *present value.* Present value means: "What is a future amount worth in today's dollars, given a discount rate?" The discount rate is just the rate you use to reflect what you could do with money in the meantime (or what it costs you to borrow, or how uncertain the future payment is).

You don't need a finance degree to use this. You need one habit: whenever someone offers you money later, ask, "What would I accept today instead?"

Here's the intuition without heavy math:

- If you can earn 5% on your money (or avoid paying 5% interest), then $100 today can become about $105 next year.
- So $105 next year is roughly equivalent to $100 today at a 5% rate.

That's present value in one breath. You're translating future money back into today's language.

Two things make this especially practical for households:

1. Your *discount rate* doesn't have to be perfect. It has to be reasonable.
2. The goal isn't precision. The goal is to avoid being fooled by timing.

If you are carrying credit card debt at 20% APR, your discount rate for spare dollars is not 5%. It's closer to 20%, because paying down that debt produces a near-guaranteed return by avoiding interest. If you have no consumer debt, a sturdy emergency fund, and you're deciding between two job offers, you might use something like 5%–8% as a rough planning rate, reflecting long-term investing potential and uncertainty.

The exact number matters less than the mindset: money later must be discounted, because waiting has a cost.

Future Value: The Other Side of the Mirror

The sibling of present value is *future value*: "If I have money today, what can it become later?"

Future value is what makes small habits feel dramatic over long stretches. It's the reason a $50/month increase in savings can matter more than an occasional windfall. It's also the reason debt is so dangerous: interest compounds against you when you carry balances.

For day-to-day household decisions, future value answers a question most people don't ask often enough:

> If we don't spend this money, what could it become?

Not in a moralizing way. In a practical way. Because $200 not spent is not merely $200 saved. It can be a month of insurance premiums in a rough patch, the difference between paying cash for a car repair and financing it, or the start of a fund that eventually eliminates a stressful monthly payment.

This is where the previous idea of opportunity cost gets sharper: you're not only choosing between *things*. You're choosing between *timelines*. You're choosing what your household will be able to do later because of what you do now.

A Real-World Case Study: The Bonus Choice That Looked Like a No-Brainer

A couple—call them Elena and Marcus—faced a common fork in the road when Marcus changed jobs. He was offered one of two compensation packages:

- Option A: $12,000 signing bonus paid immediately, with a requirement to repay it if he left within a year.
- Option B: $1,100 more per month in salary (about $13,200 per year), paid over time with no repayment clause.

At first glance, Option B looked better: higher annual dollars, and it would keep paying after the first year. The trouble was their reality. They had a thin emergency fund, a newborn, and a car that was making an ominous noise. Elena's income was about to drop during maternity leave. Their household didn't need "more money in theory." It needed cash at the exact right moment to prevent a chain reaction.

Time value turned this into a different question: *Which option reduces the odds of expensive borrowing in the next six months?*

They took Option A and used the signing bonus to:

- fully fund one month of expenses as a starter emergency cushion,
- fix the car before it failed at the worst possible time, and
- pay off a small credit card balance they'd been floating.

Did they end up with a lower annual income than Option B? Slightly, on paper. Did they end up with a stronger household balance sheet and less fragility? Yes, immediately.

And here's the part that people miss: by avoiding a crisis, they avoided the kinds of costs that don't show up in offer letters. No late fees. No high-interest balances. No panic purchases. No taking time off work at a bad moment. No "we'll just put it on the card for now." The signing bonus wasn't just money. It was a timing tool.

Time value doesn't always push you toward the immediate payout. But it forces you to respect the household context. A family with ample savings might prefer the higher monthly salary because it supports long-term contributions and is less conditional. A family with thin savings might rationally prioritize cash now to avoid the interest and stress spiral that comes from being under-resourced at exactly the wrong time.

Same options, different right answer, because time value is about *your* timeline.

The Household ``Interest Rate" You're Already Paying (Whether You Call It That or Not)

If you carry debt, you already live in a world governed by time value.

A credit card balance is the clearest example: every month you don't pay it off, the balance grows. That growth rate is the interest rate, and it represents the price of using money now and paying later.

But even households without revolving debt face implicit interest rates:

- Paying a late fee is a brutal, short-term interest rate on cash you didn't have.
- Overdraft fees are interest rates in disguise.
- Financing a purchase because you didn't have savings often costs more than the purchase itself.
- Missing an employer match because "we'll start later" is a negative return you feel years down the line.

Time value helps you see these not as isolated annoyances, but as predictable outcomes of decisions about *when* money is available.

This can feel unfair, and sometimes it is. Many of the highest "interest rates" in life fall hardest on people with the least slack. But the household-level strategy is still the same: build enough liquidity and reduce enough high-interest obligations that you stop paying premium prices for timing problems.

Why ``0% APR" Isn't Automatically Free

Retailers and lenders love selling the idea that time can be bent without cost: *buy now, pay later*. Sometimes it's genuinely a good deal. Sometimes it's a trap with polite manners.

A 0% offer is only truly free if two conditions are met:

1. You would have bought the item anyway, at the same price, without financing.
2. You will pay it off in full before the promo period ends, *without* crowding out your ability to handle emergencies.

If the financing tempts you to buy more than you planned, time value flips against you: you're using future cash flow to pay for a present lifestyle increase. If the promo ends and the interest rate jumps, the deal retroactively becomes expensive. And even if you pay it off on time, there's an opportunity cost: the monthly payments reduce your flexibility and increase fixed commitments.

In other words, 0% APR can be free in dollars and still be costly in *options*. Households don't get into trouble only because they pay interest. They get into trouble because they stack obligations until their calendar and cash flow are brittle.

The time value question to ask is: *What does this monthly payment prevent us from doing if something goes sideways?* If the answer is "it would force us to carry a balance or miss a bill," then the later option has to be *much* better to win. If "now" is secure but costs you growth, then choose growth once you can truly afford the wait.

The Three Levers That Matter Most (And the One People Overrate)

Most household financial outcomes are driven by three levers:

- *Time horizon*: how long money can work (or debt can grow) before you interrupt it.

- *Rate*: what you earn (or pay) as a percentage.
- *Ongoing contributions*: whether you keep adding to savings or keep carrying balances.

People love to obsess over a fourth lever: compounding frequency (monthly vs daily vs annually). It matters, but far less than the big three in most real household situations. A family that saves $400 per month for 20 years at a reasonable return will beat a family that saves $0 per month while hunting for a slightly better interest calculation.

Time value is not a scavenger hunt for optimization. It's a framework for making fewer decisions that you later have to "fix" with expensive effort.

How to Use Time Value Without Doing Spreadsheet Gymnastics

You can apply time value in under a minute using a few household-friendly rules.

Rule 1: Compare money on the same timeline. If one option pays now and another pays later, translate the later one into a smaller "today" number in your head. Even a rough discount improves decisions.

A quick mental shortcut:

- Over one year, discount 5%–10% for ordinary tradeoffs.
- Discount more if the payment is uncertain, delayed, or you are likely to borrow in the meantime.

Rule 2: Your highest-interest debt sets your minimum "return." If you're paying 18% on a credit card, paying it down is like earning an 18% return, nearly risk-free. That doesn't mean you ignore retirement. It means you stop treating extra payments as "not investing." Avoiding interest is investing in your future cash flow.

Rule 3: Liquidity has value, especially in two-income households. If your household only works when both incomes show up like clockwork, cash on hand is not laziness. It's risk management. A smaller return with higher liquidity can be smarter than chasing an extra percentage point while leaving yourself exposed.

Rule 4: Beware payments that turn choices into obligations. A monthly payment is not just a cost. It's a claim on your future. Stack enough claims and you don't own your income anymore; your obligations do.

An Actionable Takeaway: The ``Today or Tomorrow" Script

The next time you're choosing between a benefit now and a bigger benefit later—bonus vs salary, paying cash vs financing, paying down debt vs investing, keeping savings vs prepaying something—use this script:

> *If we take the later money, what do we have to do in the meantime? Borrow, stress, delay, or risk? If we take the money now, what future gain are we giving up, and is that trade worth the stability?*

Write the answers as plain sentences, not numbers. If "later" requires borrowing or leaves you one surprise away from a credit card balance, the later option has to be *much* better to win. If "now" is secure but costs you growth, then choose growth once you can truly afford the wait.

A household that learns to price time stops being impressed by big-looking numbers with distant dates attached. It starts valuing what money is really for: buying good decisions when life is loud.

1.3 The Silent Killer: Inflation

Inflation is the financial equivalent of a slow leak in a tire. Nothing dramatic happens on day one. You still drive to work. The steering still feels fine. Then, one morning, you're running late and the car feels heavy and wrong—and you can't remember the last time you checked the pressure.

That's how inflation works in a household. Prices don't usually double overnight. They creep. You adapt without noticing. You switch brands, skip the extras, tell yourself it's a weird month. And then, months later, you realize the same life costs more *every single month*, even though your income didn't rise at the same pace. The danger isn't only that things get expensive; it's that inflation turns yesterday's plan into today's underfunded wish.

The time value idea makes the punchline unavoidable: money has a time dimension. Inflation is what happens when time makes money smaller.

The One Sentence Definition That Actually Matters

Inflation is a broad rise in prices over time. The practical effect is brutally simple: *inflation reduces purchasing power*. Purchasing power is how much real life your dollars can buy.

You can feel this even if you never read an economic report. If your grocery bill used to be $180 and it's now $230 for the same cart, your purchasing power fell. If your rent renewal jumps while your salary stays flat, your purchasing power fell. If your kid's summer camp costs more than last year and you didn't get a raise, your purchasing power fell. Inflation is not an abstract percentage; it's the gap between your paycheck and the price tag on normal life.

And because it's broad, it hits in places you can't easily avoid. You can choose a cheaper phone plan, but you can't opt out of the fact that food, housing, insurance, and healthcare are now playing on a higher difficulty setting.

Two Inflations: The Market One and the Lifestyle One

Most people think inflation is something the economy does *to* them. That's the first inflation: *market inflation*, the general rise in prices across the economy.

The second inflation is sneakier because it feels like progress: *lifestyle inflation*. That's when your standards rise as your income rises, and the upgraded life quietly becomes the new minimum.

Lifestyle inflation doesn't require luxury. It can be very normal:

- paying for convenience because both adults are exhausted,
- adding subscriptions that start as "trial months",
- upgrading a car because reliability anxiety is real,
- moving to a pricier neighborhood for schools,
- turning occasional takeout into a default.

Market inflation makes the same lifestyle cost more. Lifestyle inflation makes you buy a bigger lifestyle. The combination is how households end up earning more over time and still feeling like they're running out of money.

The silent killer is not that prices rise. It's that your life expands to fill your income *while* prices rise. That double squeeze is how a household can appear successful and still feel financially brittle.

Nominal vs. Real: The Numbers That Lie and the Numbers That Tell the Truth

Most financial numbers you hear are *nominal*. Nominal means the raw number in dollars: your salary, your rent, your grocery bill. Nominal numbers are the ones printed on pay stubs and contracts.

But what you live on is *real* money. Real means adjusted for inflation—in other words, measured in purchasing power. Real numbers answer the only question that matters: *Is my standard of living actually improving, or am I just holding the same ground with bigger-looking numbers?*

A raise can be nominally good and really bad. If you get a 3% raise in a year when your personal cost of living rises 6%, you took a pay cut in real terms. You might not notice immediately because the paycheck is bigger. Then you notice at the checkout line and the insurance renewal and the childcare invoice.

When households plan in nominal terms, they make promises their future selves can't keep. "We'll save $500 a month for college." "We'll retire on $80,000 a year." "We'll keep our spending around $4,500." Those are not plans unless you attach them to purchasing power.

Inflation turns fixed targets into moving targets. Planning in real terms is how you stop being surprised by that.

A Quick Way to Estimate Real Return (Without Turning Into an Economist)

When your money grows, you want to know if it's actually getting ahead of inflation. That's called a *real return*. Real return is the growth of your money *after* inflation erodes purchasing power.

The approximate relationship is simple:

$$\textit{Real return} \approx \textit{Nominal return} - \textit{Inflation}$$

If a savings account pays 4% and inflation is running 3%, you're earning about 1% in real terms. If your investment portfolio returns 8% and inflation is 3%, you earned about 5% in real terms. If your wage grows 4% and your household costs rise 6%, you lost about 2% of purchasing power.

This is not about precise forecasting. It's about keeping your brain from being fooled by dollar amounts that look larger but buy less.

Your Inflation Rate Is Not the Headline Inflation Rate

Inflation numbers in the news are averages. Your household is not an average. You have a specific mix of spending: housing, childcare, commuting, insurance, healthcare, groceries. If one of those categories is a big share of your budget, your lived inflation can be much higher than whatever number flashes across the screen.

A family paying for full-time childcare experiences inflation differently than a retired couple. A renter in a hot market experiences inflation differently than a homeowner with a fixed-rate mortgage. A household with chronic medical costs experiences inflation differently than a household with none. This is why arguments about "real inflation" get so heated: people aren't lying; they're living in different baskets.

Instead of debating the headline, track your own reality. Your bank statements already contain your personal inflation data. If your household's baseline costs have climbed 10% over two years, that's your inflation problem, even if the national average is lower. You don't pay bills with national averages.

A Case Study: The Raise That Didn't Change Anything

Consider Jordan and Sam, two incomes, one toddler, one aging car, and the kind of monthly budget that looks stable until you zoom in.

Jordan got a raise: from $92,000 to $98,000. Six thousand dollars feels meaningful. Their friends congratulated them. The household felt, briefly, like they'd taken a step forward.

Three months later, they were back to the same tight feeling. Here's what happened, in pieces that felt unrelated at the time:

- Their rent renewal increased by $225 per month.
- Their childcare tuition increased by $150 per month.
- Their auto insurance premium jumped at renewal.
- Groceries crept up by $40–$60 per week, depending on the week.
- The car needed tires, and they financed part of it because the checking account didn't have room.

None of these line items was a moral failing. It was life, plus market inflation, plus a budget without enough slack.

The nominal raise was real. The real raise—the increase in their purchasing power—was close to zero, and in some months negative. Their household wasn't being dramatic when they said, "It doesn't feel like more money." It literally wasn't, once the cost of maintaining normal life rose.

The fix wasn't "spend less on lattes." Their budget didn't have a latte problem; it had a baseline-cost problem. The household needed two things:

1. a system for tracking baseline spending in real terms, and

2. an intentional stance against lifestyle inflation so raises could rebuild slack instead of disappearing into higher commitments.

Once they saw the pattern, they made two changes that mattered more than any couponing strategy: they negotiated a longer lease to reduce rent volatility and redirected the entire net raise (after taxes) into a combination of emergency savings and paying down their highest-interest debt. They didn't feel rich. They felt *less fragile*. That's the real win inflation tries to steal.

The Place Inflation Hurts Most: Fixed Costs

Inflation is annoying everywhere, but it becomes dangerous when it pushes up *fixed costs*. Fixed costs are the bills that show up whether you're having a good month or a bad one: rent or mortgage, car payments, insurance, childcare, minimum debt payments, subscriptions you forgot you had.

Variable costs can be trimmed. You can cook more, pause travel, reduce entertainment, delay purchases. Fixed costs are less flexible. They turn a temporary income dip into a crisis.

Here's the trap: in response to inflation stress, households often finance more of life. Payments feel manageable in the moment. Payments are also how you turn a short-term problem into a long-term obligation. Inflation nudges you toward higher baseline costs; financing locks those costs into your future.

This is why the most inflation-resistant households are not necessarily the highest-income ones. They're the households with lower fixed costs relative to income and enough cash to absorb bumps without converting them into debt.

Why Inflation Makes People Feel ``Bad at Money''

Inflation creates a psychological tax. When prices rise, the same spending suddenly feels irresponsible. You buy the same groceries and feel guilty. You renew the same insurance and feel like you failed. You look at the same budget and wonder why you can't make it behave.

This is where households start fighting with each other. One partner thinks the other is overspending. The other feels blamed for costs they can't control. The tension is real because the squeeze is real.

A useful reframe: when inflation is the driver, the goal isn't to find a villain in the household. The goal is to update the household model of reality. Old budgets often assume old prices. Old goals assume old costs. Inflation forces a re-pricing of life.

If your household updates the model together, money talks become calmer because you stop arguing about character and start solving for math and tradeoffs.

Planning in Constant Dollars: Making Goals Inflation-Proof

A powerful trick used in professional finance is to think in *constant dollars*. Constant dollars are dollars adjusted for inflation, so you're comparing purchasing power across time.

Households can borrow this idea in a simple way: when you set a goal that's more than a year away, attach an inflation assumption and update the target annually.

For example:

- If you want $20,000 in an emergency fund, ask: $20,000 in *today's* dollars or five years from now?
- If you plan to spend $6,000 on a vacation in three years, ask: will that vacation cost $6,000 in three years, or is that today's price?

Even a rough inflation assumption (say 3% annually) changes the plan. A $20,000 goal in today's dollars becomes about $23,000 in five years at 3% inflation. The household that ignores inflation feels perpetually behind. The household that builds it in feels realistic and steady.

This is especially important for long goals like retirement and college. The dollars you will need decades from now are not the dollars you need today.

Lifestyle Creep: The Inflation You Can Actually Control

Market inflation isn't negotiable. Lifestyle inflation is.

Lifestyle creep happens when every raise is pre-spent in your imagination. The new income gets mentally assigned to nicer groceries, better vacations, upgraded cars, bigger housing, more conveniences. There's nothing wrong with enjoying money. The issue is when enjoyment becomes a permanent expansion of fixed costs, because fixed costs are what turn a household into a machine that must keep running at maximum output.

A simple rule that works surprisingly well in real life: treat raises like a split decision until proven otherwise.

For example, if your household gets a $500/month increase in take-home pay, decide ahead of time:

- how much buys present-life improvements (guilt-free), and
- how much buys future freedom (savings, debt payoff, sinking funds).

This is not about austerity. It's about making sure inflation doesn't eat the raise and lifestyle creep doesn't bury it. The raise becomes an engine for resilience, not just a temporary feeling.

The Most Practical Inflation Defense: Sinking Funds

A sinking fund is an unglamorous term for a beautiful thing: money you set aside every month for a known future expense. It's how you stop being surprised by expenses you already know are coming.

Sinking funds are inflation's enemy because they convert chaos into planning. When prices rise, households with sinking funds aren't immune, but they're less likely to reach for debt.

Common sinking funds:

- car repairs and replacement,
- medical out-of-pocket costs,
- home maintenance,
- kids' activities,
- travel and holidays.

Inflation makes these categories more expensive over time. A sinking fund that gets updated annually is the household equivalent of checking tire pressure. It's boring. It prevents blowouts.

Actionable Takeaway: Run a ``Real Life Price Check" Once a Quarter

Once per quarter, spend 20 minutes doing this with your partner (or solo, if you manage money alone):

1. Pick five baseline categories: housing, groceries, insurance, childcare/school costs, and transportation.
2. Compare your average monthly spending in each category to the same quarter one year ago.
3. Circle the categories that rose faster than your take-home pay.
4. For each circled category, choose one response:
    - *Reduce*: change behavior, renegotiate, shop around.
    - *Replace*: find a lower-cost substitute that still fits your values.
    - *Reprice*: accept the new reality and raise the budget target so you stop pretending it's temporary.

Inflation thrives on vague awareness and delayed action. This small ritual turns inflation into something you can see, name, and respond to—before it quietly rewrites your entire financial life in the background.

1.4 The Engine of Wealth: Compound Growth

If you've ever planted something—a tomato, a houseplant, a stubborn patch of grass—you know the most annoying part: for a while, nothing seems to happen. You water. You wait. You doubt your competence. Then one day you notice new leaves and realize growth has been happening the whole time; it just wasn't dramatic enough to satisfy your brain.

Compound growth has the same problem. It's one of the most powerful forces in household finance and one of the easiest to abandon, because early results look like a rounding error. The danger is not that compounding is complicated. The danger is that compounding is *patient*, and most of us are not.

Compounding means you earn returns on your original money *and* on the returns that money already earned. Interest on interest. Growth on growth. The snowball doesn't just roll; it gathers more snow, which makes it better at gathering more snow.

That's the engine. And like any engine, it doesn't care whether it's pushing you forward or dragging you under. Compounding builds wealth in investments, but it also builds misery in debt. The math is neutral. Your choices decide which side you experience.

The ``Boring Phase" Is Not a Bug---It's the Price of Admission

People love the *idea* of compounding and hate the calendar it requires.

In the early years, most of the growth comes from your contributions, not from returns. Your account balance rises because you put money in, not because the money magically multiplied. That's why it feels slow. Your brain wants fireworks; compounding starts with drip irrigation.

This is the emotional trap where many households quit:

- "We've been saving for a year and it doesn't feel different."
- "The market went down—what's the point?"
- "We'll start investing when we have more."
- "We'll catch up later."

Later is expensive. Not because you're punished for waiting, but because you lose the one ingredient you can never buy: *time*. Compounding is the closest thing finance has to a cheat code, and the cheat is simple: start earlier than you feel ready.

A Quick Story: The Two Raises That Changed Nothing (Until They Did)

Consider a couple, Nina and Andre. They weren't big earners at first: early 30s, one kid, renting, one student loan still hanging around. They wanted to invest but felt embarrassed by how small their numbers were. The idea of putting $150 a month into a retirement account felt like trying to fill a swimming pool with an eyedropper.

Still, they automated it. $150 a month into a Roth IRA for Nina. Andre put 4% into his 401(k) to capture the employer match. They set it up and tried not to overthink it.

Year one was underwhelming. They contributed, the balance moved, the market wobbled, and nothing about their daily life felt improved. But something important happened under the surface: they built a system that made saving the default instead of a heroic act.

Then two raises came, not huge, but steady. Instead of inflating their lifestyle by the full amount, they split the difference: half to present life, half to their automated contributions. $150 became $250, then $350. Andre nudged his 401(k) contribution to 6%, then 8%. Still not glamorous. Still not Instagram-worthy.

A decade later, the story sounded different. Their investments didn't just reflect what they had contributed; the accounts had gained a second job. Growth had started contributing too. The balances had reached the point where a normal market year could add more than they were adding themselves. That's the moment compounding becomes visible. The engine finally makes noise.

Their secret wasn't stock-picking or perfect timing. It was consistency through the boring years and the discipline to let raises buy future freedom, not just more monthly commitments.

The Rule of 72: A Mental Shortcut That Changes How You See Rates

You don't need complex formulas to develop good instincts. One of the simplest tools is the *Rule of 72*. It's a quick way to estimate how long it takes money to double at a given annual rate.

Doubling time (in years) $\approx 72 \div$ rate

A few examples:

- At 6% growth: $72 \div 6 \approx 12$ years to double.
- At 8% growth: $72 \div 8 = 9$ years to double.
- At 3% growth: $72 \div 3 = 24$ years to double.

This does two useful things for a household. First, it makes rates feel real. "Eight percent" stops being an abstract number and starts being "money doubles about every nine years." Second, it highlights why high-interest debt is dangerous: a credit card charging 24% can double what you owe in roughly three years if you let it run.

Compounding is a power tool. You don't want it pointed at your ankle.

The Mechanics You Actually Need: Contributions Matter More Than Cleverness

Most people overestimate how much they need to know to benefit from compound growth.

To harness compounding, the biggest drivers are:

- *Start date*: earlier is disproportionately better.
- *Contribution rate*: what you add consistently.
- *Time in the market*: staying invested long enough for the engine to build momentum.

Notice what isn't on that list: perfect predictions. Households can waste years waiting for "the right time." Meanwhile, time passes. The market will always give you reasons to hesitate. There will always be scary headlines. There will always be someone online claiming disaster or promising easy returns.

Compounding rewards the household that keeps showing up.

That doesn't mean you ignore risk. It means you separate two decisions:

1. *The decision to invest consistently*, which builds the habit and the timeline.
2. *The decision of what to invest in*, which should be boring, diversified, and appropriate for your risk tolerance.

The first decision is often the one that makes or breaks outcomes. The second decision matters too, but it rarely compensates for a decade of delay.

Compounding's Evil Twin: Debt That Grows While You Sleep

If compound growth is the engine of wealth, compound interest on debt is the engine of entrapment.

Credit cards are the clearest example because they combine three features that are brutal together:

- high interest rates,
- compounding (interest on interest when balances carry),
- minimum payments designed to keep you paying for a long time.

The minimum payment feels like relief. In reality, it's a treadmill fee. It keeps you current while the balance quietly stays alive.

This is why households feel like they're "working so hard and not getting ahead." If compounding is happening on both sides of your life—investments growing slowly while high-interest debt grows quickly—the net effect can feel like running uphill with a backpack.

The most powerful move is not always maximizing returns. It's stopping negative compounding. Paying off high-interest debt is like turning off a leak that drains your future every month.

A Concrete Example: The Same $300, Two Futures

Imagine a household has $300 per month of "extra" cash flow. Not imaginary extra—real extra after bills.

They have two paths:

Path A: Invest $300/month. Assume a long-term average return of 7% annually (not a promise, just a planning assumption). Over 30 years, consistent contributions plus compounding can build a serious balance.

Path B: Carry $10,000 on a credit card at 22% APR while investing $0. The household pays the minimums, occasionally throws a little extra at it, but the balance lingers. Interest compounds fast, and the debt becomes a permanent background payment.

Here's what's sobering: the difference between these futures may not be investing brilliance. It may be simply eliminating the debt compounding against them so that their $300 can finally compound *for* them.

Many households attempt Path A while living in Path B. They contribute a little to retirement while also paying heavy interest. That can still be reasonable in some cases (especially to capture an employer match), but emotionally it often feels like pushing the gas and the brake at the same time.

A more coherent strategy is usually:

- capture any "free money" like a match if available, then
- attack the highest-interest debt aggressively, then
- increase investing once the negative compounding is shut down.

Coherence matters because it keeps you motivated. Progress you can feel is progress you'll repeat.

The Household Trick That Makes Compounding Automatic: ``Pay Yourself First"

Compounding needs consistency. Consistency is hard when saving relies on willpower at the end of the month.

The most reliable method is simple: *pay yourself first*. That means your saving or investing happens automatically when income arrives, before the money becomes available for everything else.

This is less about discipline and more about physics. Money that sits in checking becomes spendable. Money that moves automatically into a retirement plan or brokerage account becomes less tempting to "borrow" for a weekend or a convenience purchase.

The key is to start with an amount that won't cause a bounce. It's better to automate $50 and increase it quarterly than to attempt $500, fail, and decide you're not an investing person.

Compounding doesn't require heroics. It requires a system that works during normal, boring weeks when nobody is feeling inspired.

The Emotional Side: Why Compounding Feels Like It ``Suddenly'' Works

People often say, "It felt like nothing happened, and then it took off." That's not magic. That's the math of exponential growth colliding with human perception.

Early on, returns are small because the base is small. Earning 7% on $2,000 is $140. That's not life-changing. Earning 7% on $200,000 is $14,000. That is very noticeable. Same rate, different base.

This is why the early years are the hardest: the reward is quiet while the sacrifice is loud. You feel the $300 leaving your account. You don't feel the $14 in earnings. It's psychologically backwards.

A helpful mindset shift is to treat early contributions as purchasing a future engine. You're not saving for the sake of saving; you're buying an asset that can eventually carry part of your load.

Make It Real: A ``Compounding Calendar'' Instead of a Goal Number

Big goal numbers can feel like fantasy. "We need $1.2 million." "We need to save $20,000 a year." Those statements can be accurate and still paralyzing.

A more motivating approach is to focus on a compounding calendar:

- What can we automate *this month*?
- What can we increase by *1%* when we get the next raise?
- What debt can we eliminate in the next *six months* to free cash flow for investing?

Compounding is built out of boring repeats. The calendar keeps your attention on repeats you can actually execute.

The Chart That Explains Why Patience Gets Paid

At first, linear saving and compounding don't look that different. Then the gap opens. The end of the timeline is where the curve gets dramatic—not because the saver suddenly became smarter, but because time finally had enough material to work with.

The most dangerous takeaway is: "I'll start later, because the last ten years are where it really grows." That's like saying you'll plant the tree later because the shade is best when it's big. The last ten years grow because the first twenty exist.

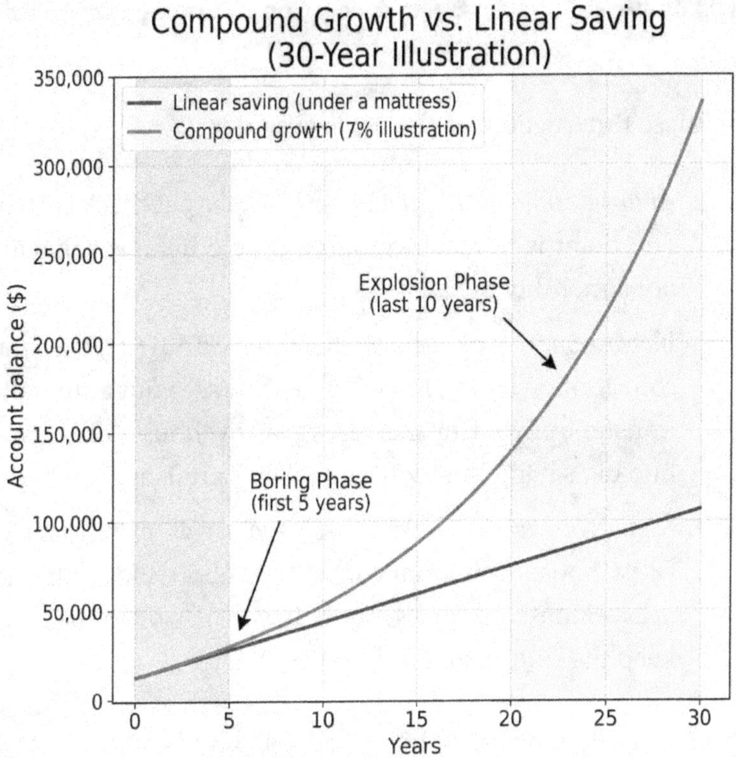

A chart like this can be inspiring or depressing depending on what you do with it. The inspiration comes from realizing that consistent habits can become surprisingly large outcomes. The depression comes from realizing time passes either way.

The most useful emotion is neither. It's resolve. A household doesn't need perfect conditions to begin compounding. It needs a decision to stop waiting for a version of life where saving feels effortless.

Actionable Takeaway: Build Your ``Compounding Stack" This Week

Do these three actions in the next seven days:

1. *Automate one contribution.* Even $25 per paycheck counts. The point is to make compounding a default behavior, not a monthly debate.
2. *Increase one rate by one step.* If you have a 401(k), raise the contribution by 1%. If you don't, increase your automatic transfer by $25. Tiny increases are how households grow into big saving rates without feeling punished.
3. *Name your highest-interest debt and set an extra-payment rule.* Choose an amount you will pay above the minimum every month until it's gone. If you get overwhelmed, keep the rule small but non-negotiable.

That combination—automatic investing, gradual increases, and shutting down negative compounding—is how the engine of wealth becomes something you can actually feel. Not tomorrow, not next month, but reliably, year after year, until the math starts doing the heavy lifting for you.

Chapter 2

Designing Your Household Operating System

Most couples run their finances like two roommates frantically trying to split a pizza bill in the dark. But a resilient household shouldn't rely on guesswork or willpower; it needs an operating system. By borrowing a few lightweight strategies from the business world-like a clear balance sheet, a standardized 'monthly nut,' and conflict-free communication protocols-you can build a background machine that handles the stress for you. This chapter isn't about turning your marriage into a corporation; it's about automating the boring stuff so you can get back to being a family.

2.1 The Household Balance Sheet: Your Financial Scoreboard

On paper, Maya and Chris looked unstoppable. Two solid salaries, health insurance through work, a house in a good school district, and the kind of calendar that required color-coding just to stay polite to friends. They weren't reckless. They weren't impulsive. They were, in the way most people mean it, *responsible*.

Then the transmission failed.

Not a dramatic catastrophe—no hospital stay, no job loss, no natural disaster. Just a $3,800 repair on a Tuesday. The real surprise wasn't the number; it was the scramble. They had retirement accounts, home equity, and a respectable income. They also had $612 in their checking account, and the credit cards were already carrying a balance from last month's "we'll catch up next paycheck" logic.

That's the moment you learn the difference between *looking* financially healthy and *being* financially resilient.

Income is a speedometer. It tells you how fast money is arriving. A household balance sheet is a scoreboard. It tells you who's actually winning the game.

Why income lies (or at least, tells a half-truth)

A high income can be a financial life jacket. It can also be a trapdoor.

If every dollar of earnings is pre-committed—mortgage, car payments, childcare, subscriptions, minimum debt payments, the quiet creep of "we deserve this" spending—your income becomes a treadmill. You run hard, you sweat, and the scenery never changes. Worse: treadmills are fragile. A single stumble (a medical bill, a layoff, a broken HVAC) can throw you off.

Net worth is what's left after you add up what you own and subtract what you owe. It's a snapshot, not a monthly story. And that's exactly why it's powerful. Cash flow shows motion; net worth shows position. When your net worth is trending upward in a steady, boring way, you're building gravity in your favor.

The most common mistake households make is believing financial strength is primarily about earning more. Earning more helps—but the scoreboard doesn't care about effort. It only cares about results.

The balance sheet in plain English

A household balance sheet is a list with two sides:

- *Assets*: things you own that have value.
- *Liabilities*: money you owe.

Then:

$$Net\ Worth = Assets - Liabilities$$

It's simple enough to say in one breath. The difficulty is emotional. Because the balance sheet does not negotiate. It doesn't care that the payment is "only" $489. It doesn't care that you got a raise. It doesn't care that you *intend* to pay off the card.

It just tallies.

That's also why it's such a relief. A good scoreboard turns vague anxiety into specific math. Specific math can be improved.

A crucial twist: not all assets are created equal

Here's where families get tripped up: they confuse *valuable* with *useful in a crisis*.

Yes, your home has value. Yes, your retirement account has value. But if your kid breaks an arm tomorrow, the hospital won't accept "approximately $240,000 in home equity" as payment. They want money that can actually move.

So when you build your balance sheet, sort your assets into two buckets:

- *Liquid assets*: cash (and cash-like things) you can use quickly without penalties or drama.
- *Illiquid assets*: value that exists, but is harder to access (or costly to access) when you need it.

Liquidity is the difference between a bad week and a financial spiral.

A house is an asset. But turning that value into spending power usually requires a sale, a refinance, or a home equity loan/line—each with time, paperwork, and the possibility that the market mood has changed since last year. Retirement accounts are assets, but early withdrawals can trigger taxes and penalties. Even a paid-off car is an asset, but selling it might solve the bill and create a transportation problem.

Liquid assets are boring. That's their superpower.

The difference between ``rich on paper'' and ``safe in real life''

A household can have a high net worth and still be financially brittle. This is the classic "house-rich, cash-poor" pattern.

Imagine two households with the same net worth:

- Household A: $300,000 home equity, $20,000 in retirement, $1,500 in cash, and $25,000 in credit card and auto debt.

- Household B: $80,000 home equity, $70,000 in retirement, $25,000 in cash, and $0 in consumer debt.

They might have similar net worth totals, but they live in different emotional climates. Household A feels like they're always one surprise away from a problem. Household B has oxygen. Not because they're "better people," but because their assets are positioned for flexibility.

This is why the scoreboard needs two numbers, not one:

- *Net worth* (the overall score)
- *Liquid net worth* (how much is available without selling your life)

A practical definition:

$$\textit{Liquid Net Worth} = \textit{Liquid Assets} - \textit{Short-term Liabilities}$$

Short-term liabilities are what can hurt you fast: credit cards, medical bills on payment plans, personal loans, anything with a payment due next month. (Your mortgage is real, but it's not usually "call a collections agency tomorrow" real if you're one day late.)

Liquid net worth is the number that buys you sleep.

A concrete household example: building a real balance sheet

Here's a simplified balance sheet for Maya and Chris, the couple with the broken transmission. The numbers are rounded, but the pattern is painfully common.

Category	*Item*	*Amount*
Liquid Assets	Checking + savings	$1,200
	HSA balance	$2,600
Illiquid / Long-term Assets	Retirement accounts (401(k)/IRA)	$74,000
	Home (estimated market value)	$520,000
	Vehicles (resale value)	$22,000
Liabilities	Credit cards	$14,800
	Auto loan	$18,500
	Student loans	$27,000
	Mortgage balance	$412,000
Totals	Total assets	$619,800
	Total liabilities	$472,300
Net Worth	Assets – liabilities	$147,500

If you stop there, this household looks fine: positive net worth, retirement contributions, a home that has likely appreciated. But the transmission didn't ask for their net worth. It asked for cash flow and liquidity.

Now look at liquid net worth:

- Liquid assets: $1,200 + $2,600 = $3,800
- Short-term liabilities (credit cards): $14,800

Liquid net worth: $3,800 – $14,800 = −$11,000.

That negative liquid net worth is why a mid-sized car repair felt like a crisis. Their wealth was real, but it was stored behind glass.

The ``true asset" test: does it put money in your pocket?

A helpful way to reduce balance-sheet confusion is to ask a blunt question:

> If I keep this thing, does it tend to put money in my pocket, or does it tend to take money out?

Some items are obvious:

- A savings account pays a little interest and holds spending power: it puts money in your pocket (slowly).
- A credit card balance takes money out every month: it takes money out.

Other items are emotionally messy:

- A primary home usually *feels* like an asset. Financially, it can be both: it may appreciate, but it also creates ongoing costs (taxes, insurance, repairs) and can be hard to monetize quickly. Call it an asset on the balance sheet, but don't treat it like an emergency fund.
- A car is often a liability disguised as an asset. It has resale value, yes, but it depreciates and demands maintenance. On the balance sheet it's an asset; in your monthly life it behaves like a mouth to feed.

This isn't about guilt. It's about clarity. When you know which parts of your "wealth" are actually available, you stop making plans that require miracles.

The quiet villains: liabilities that don't feel like debt

Most families can list their mortgage and student loans. The stealth liabilities are the ones that live in autopay limbo or emotional denial.

Common examples:

- *Buy now, pay later* plans that feel like "just a small installment."
- *Medical bills* on a payment plan (especially when they're scattered across providers).
- *Taxes owed* from under-withholding or 1099 income (a liability even before the IRS sends a letter).
- *Deferred maintenance* (not a line item you owe to a bank, but a real claim on future cash).

Deferred maintenance deserves special respect. A roof doesn't care about your budget. It cares about physics and time. If you own a home, you effectively have a silent loan from your future self. Your balance sheet won't show it, but your resilience will.

A practical workaround: create a "house reserve" sinking fund or at least note "expected big repairs" as a shadow liability in your personal notes. The point isn't precision; it's honesty.

Net worth is a photo, not a movie

A balance sheet is taken on a date. Pick a day—the first of the month, your payday, the day after bills clear—and treat it like a snapshot.

Why snapshots work:

- They stop you from arguing with yourself. "I feel like we're doing better" becomes "we are up $2,400 since October."
- They make progress visible even when life is expensive. Paying down a card by $500 might not change your lifestyle, but it changes your score.
- They create a feedback loop. People improve what they can see.

Also: don't demand perfection. Your home value is an estimate. Your car value is a guess. That's fine. The goal is trend accuracy, not courtroom accuracy.

How to make the scoreboard actually change

Net worth improves in only a few ways:

- Assets go up (saving, investing, paying down a mortgage increases equity, buying something that holds value).
- Liabilities go down (debt payoff).
- Or both, at the same time (the best kind).

The emotional trap is thinking you need a dramatic move. You don't. You need a repeatable one.

For most households, the highest-leverage plays are unglamorous:

- Building liquid assets until "random Tuesday expenses" become annoying instead of dangerous.

- Eliminating high-interest consumer debt so your money stops leaking out in interest.
- Protecting your upside (retirement contributions, especially if there's an employer match).

If you want your net worth to rise faster than your stress, prioritize *liquidity first, leakage second, long-term growth always.* In other words: get some cash under you, plug the holes, keep planting seeds.

A quick gut-check: the three numbers worth knowing

If you only tracked three balance-sheet numbers, make them these:

- *Net worth*: your overall score.
- *Liquid net worth*: your ability to handle trouble without chaos.
- *Consumer debt*: credit cards, personal loans, anything that charges you for yesterday.

Each number tells a different truth. Net worth tells you whether you're building wealth. Liquid net worth tells you whether a surprise becomes a crisis. Consumer debt tells you how much of your future paycheck is already spent.

Actionable takeaway: the 30-minute scoreboard sprint

Set a timer for 30 minutes and build your first household balance sheet with "good enough" numbers.

1. Write down every account with money in it (checking, savings, retirement, HSA). Use today's balances.
2. Estimate the big assets (home value, car value). Use a conservative number you wouldn't be embarrassed to accept in a quick sale.
3. List every debt with a balance (mortgage, auto, student loans, credit cards, medical bills, BNPL).
4. Do the subtraction: assets minus liabilities.
5. Do it again, but only with liquid assets minus short-term liabilities. Circle that number.

Then pick one concrete move that improves the circled number within the next 14 days: transfer $300 into savings, set up an automatic $50 weekly cushion, or make a targeted extra payment on the highest-interest card. The goal is not a beautiful spreadsheet. The goal is a scoreboard that starts moving in the right direction—and a household that can absorb a random Tuesday without holding its breath.

2.2 Mastering Cash Flow: Knowing Your 'Monthly Nut'

The most dangerous sentence in household finance is: *"We make good money, so we should be fine."*

Plenty of families *are* fine on paper—until the month turns into a tightrope. A kid needs braces. The water heater dies. Your hours get cut for six weeks. Nothing about those events is rare. What's rare is a household that can absorb them without panic, arguing, or a credit card swipe that starts a long, slow debt story.

Cash flow is the part of your finances that interacts with real life. It's the money that shows up, the money that leaves, and the timing mismatch between the two. When cash flow is under control, your household feels like it has power steering. When it's not, you can be earning more than you ever have and still feel like you're grinding gears.

The tool that makes cash flow understandable is your *monthly nut*: the minimum amount of money your household must spend each month to keep the lights on and avoid financial damage. It's not a "nice" month. It's not a month where you buy new shoes and take weekend trips and decide you deserve appetizers. It's the baseline operating cost of your life.

If you know your monthly nut, you can answer the questions that actually matter:

- How long could we survive on one income?
- If one of us lost a job, what expenses would we cut *first*—without negotiating in a crisis?
- Are we one surprise away from missing a payment?

That's not budgeting as a personality test. That's budgeting as a safety system.

The cash-flow problem nobody wants to admit: timing

Households don't usually fail because the math is impossible. They fail because the *timing* is brutal.

Rent or the mortgage is due on the 1st. Daycare is due weekly. Car insurance hits twice a year in one sharp bite. A property tax bill arrives like an uninvited guest. And your income? It arrives on whatever schedule your employer feels like using, which may or may not match your bills.

That timing mismatch creates a weird psychological fog. You can look at your annual salary and feel secure, then look at your checking account on the 27th and feel broke. Both feelings are honest. One is just more useful than the other.

A cash-flow system is what turns a high-income household from "why are we always behind?" into "we know exactly what's happening."

Fixed costs vs. variable costs: the difference between a cage and a dial

The core move is separating expenses into two categories:

- *Fixed costs*: obligations that are hard to change quickly. Think: rent/mortgage, minimum debt payments, insurance, childcare, basic utilities, student loans, required prescriptions.
- *Variable costs*: spending you can dial up or down without calling a bank. Think: groceries (within reason), gas, dining out, shopping, subscriptions, entertainment, gifts, travel, the "random Target run" category that somehow becomes a monthly line item.

Fixed costs are the financial equivalent of concrete. Once they're poured, you can't easily reshape them. Variable costs are more like clay. You can squish them when you need to, and in a healthy system you *plan* to squish them before you touch the essentials.

A lot of couples get stuck because they treat everything as variable-until it's not. They think, *"We'll just tighten up if we need to."* Then the moment arrives, and the "tightening" has to be huge, immediate, and emotionally painful. That's when money fights start sounding like character critiques.

The monthly nut eliminates that drama. It says: here is the floor. Here is the minimum. We can argue about the ceiling later.

The Water Tank model: why savings should be overflow, not a willpower contest

It helps to picture your household like a set of tanks that fill in order. Income pours in. It hits the non-negotiables first. If there's anything left, it spills into flexible spending. If there's still more, it spills into savings.

The reason this matters is simple: many households try to "save first" in theory, but in practice they save last—whatever happens to be left at the end of the month. And somehow, miraculously, there is never anything left.

When you accept the tank model, you stop moralizing about it. You start engineering it. You make fixed costs smaller so the first tank doesn't eat everything. You put guardrails on variable spending so it doesn't expand to fill the whole container. You build savings as a deliberate overflow path, not a guilt trip.

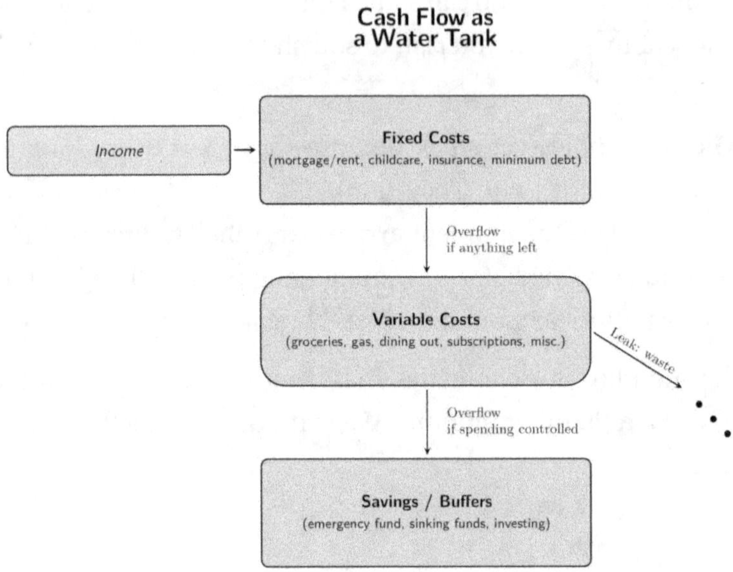

The punchline of this diagram is not "save more." The punchline is: *if the fixed-cost tank is too big, everything else becomes a fight.* If the variable-cost tank has a leak, you'll feel broke no matter how much pours in. If you want savings to happen consistently, your system has to make it the natural overflow.

Defining the monthly nut without turning your life into a spreadsheet religion

A monthly nut is the cost of survival *with dignity*. It includes the bills that keep you housed, insured, employed, and medically stable. It includes feeding the household at a basic level. It includes the minimum payments that keep your credit from catching fire.

It does *not* include "we always do this," if "this" is optional. Many households treat convenience as oxygen and then wonder why they can't breathe financially.

Here's a sane way to calculate it:

1. Pull the last 2–3 months of statements (checking, credit cards, loan payments).
2. Label each expense as fixed or variable.
3. For the monthly nut, include:
    - Housing (rent/mortgage + required HOA)
    - Basic utilities (electric, water, trash, basic internet/-phone if required for work/school)
    - Insurance (health, auto, home/renters)
    - Childcare and required school costs
    - Transportation to earn income (gas/transit + minimum car payment if you can't change it quickly)
    - Minimum debt payments
    - Groceries at a "no fancy stuff" level
    - Required medical costs/prescriptions
4. Exclude for now:
    - Dining out, alcohol, entertainment
    - Shopping beyond replacements
    - Travel, gifts beyond essentials
    - Subscriptions you can cancel in 10 minutes

Two rules keep this from becoming fantasy:

- Use real numbers, not hopes. If groceries are $900 lately, don't write $500 because it feels disciplined.
- Add a "messy life" buffer. Even a tight baseline has randomness: a school fee, a tire patch, a co-pay. A small cushion makes the nut realistic instead of brittle.

A real-world case study: the family that thought they needed a raise

Talia and Rob (two kids, one dog, one minivan with a mysterious noise) were sure they had an income problem. Their combined take-home pay was about $8,200/month, and they were still leaning on credit cards. Their conclusion was immediate: *we need to earn more.*

When they mapped cash flow, the story changed.

Their fixed costs were massive:

- Mortgage + taxes + insurance: $3,050
- Childcare: $1,700
- Car payments (two): $980
- Student loan minimums: $620
- Insurance (auto + umbrella): $290
- Phones + internet: $210
- Minimum credit card payments: $310

Fixed costs subtotal: $7,160/month

That left $1,040 for everything else: groceries, gas, kids' needs, prescriptions, school stuff, birthdays, haircuts, and the ordinary friction of living. No wonder they were using credit. Their "variable" spending wasn't indulgent; it was attempting to cover a life with what was left over.

Their monthly nut (including basic groceries, gas, and a small buffer) was closer to $7,800. That meant their household had only about $400/month of breathing room on a normal month—before any irregular expenses.

They didn't need a raise first. They needed to shrink the concrete.

The most effective move wasn't extreme deprivation; it was one structural change. They sold one of the newer cars, bought a reliable used replacement, and freed up $360/month in payment plus lowered insurance slightly. Then they called their internet/phone providers and cut $65/month. Those two moves didn't solve everything, but they changed the feel of the household. Suddenly there was room to pay cash for "random" expenses instead of financing them.

The weird thing about cash flow is that small improvements compound emotionally. Once you're no longer juggling bills, you make better decisions. You stop paying late fees. You stop using credit cards as a bridge. The bridge is where debt begins.

The minimum monthly nut vs. the comfortable monthly nut

Households often argue because one person is picturing a minimum baseline and the other is picturing a comfortable baseline.

Both are legitimate. Confusing them is the problem.

- *Minimum monthly nut*: the survival baseline, the "what we must cover" number.
- *Comfortable monthly nut*: the baseline plus the quality-of-life items that keep the household from feeling punished (a modest fun budget, a couple of subscriptions you genuinely use, slightly better groceries).

The minimum nut is for safety planning: "Could we survive a job loss?" The comfortable nut is for designing a life you can maintain without rebelling against your own budget.

A household that only tracks the comfortable number may feel fine—until it isn't. A household that only tracks the minimum number may feel like they're constantly on alert. Knowing both gives you a thermostat instead of a panic button.

The ``paycheck illusion": why some households feel broke on high income

When income is high, the temptation is to match your lifestyle to your paycheck without noticing you've done it. The first upgrade feels rational. The second feels deserved. The third arrives disguised as "normal."

Then the household becomes dependent on two incomes not because the family is extravagant, but because the fixed costs have quietly been engineered to require maximum output at all times. That's the two-income trap in its most practical form: your life becomes a machine that needs constant fuel.

The monthly nut is the antidote because it exposes dependency. If your nut is 85–90% of your take-home pay, you are living in a narrow hallway. You can still be happy there. You just shouldn't pretend it's spacious.

Turning the monthly nut into a decision-making tool

Once you have the number, use it to make decisions that are otherwise fuzzy:

Job risk decisions. If you know the minimum nut is $6,400/month and one income reliably covers $7,000 after taxes, a job loss is scary but not catastrophic. If one income covers $4,200, you know exactly how quickly you need to cut, earn, or draw on savings.

Emergency fund targets that aren't generic. "Three to six months of expenses" is only helpful if you know which expenses count. If your minimum nut is $6,400, three months is $19,200. That's not a slogan; it's a target.

Debt payoff realism. If your nut is so high there's no margin, aggressive debt payoff plans become a monthly disappointment. That's not a willpower issue. It's a math-and-structure issue. The goal becomes: lower fixed costs, stop new debt, then accelerate payoff.

Lifestyle upgrades with eyes open. Before you add a car payment, a bigger house, or a private school tuition bill, you can ask: does this raise the nut in a way that makes us fragile? You can decide yes. You just won't be surprised later.

Actionable takeaway: calculate your nut and stress-test it in one sitting

Block 60 minutes, put on a timer, and do this with whatever tools you have (bank app, spreadsheet, or a legal pad):

1. Write "*Minimum Monthly Nut*" at the top of a page.
2. List your fixed costs first (housing, childcare, insurance, minimum debt payments, required utilities/phones/internet).
3. Add a realistic baseline for groceries, gas/transportation, and required medical costs.
4. Add a small "messy life" buffer (even $150–$300) so the number can survive reality.
5. Total it.
6. Stress-test it with one question: *If one income disappeared for 60 days, what would we cut in week one?* Write those cuts down now, while nobody is panicking.

The point of the monthly nut isn't to make life smaller. It's to make your life *less breakable*. When you know the minimum cost to run your household, you stop guessing, stop negotiating under pressure, and start building the kind of margin that turns emergencies into inconveniences.

2.3 Communication and Money Meetings: The ``No-Surprise" Rule

The fight usually isn't about the $47.

It's about what the $47 *means*. It's about one partner seeing "a small treat after a brutal week," and the other partner seeing "proof we're not taking this seriously." It's about two people staring at the same bank account and reading two completely different stories—stories built from childhood, stress, pride, and the tiny humiliations of being an adult who sometimes doesn't feel in control.

If you've ever had a money argument that felt out of proportion to the purchase, you're not bad at math. You're dealing with the fact that money is a loaded subject *and* a shared resource. That combination turns ambiguity into friction. The fix isn't more spreadsheets. The fix is a household communication system that prevents financial surprises from becoming emotional landmines.

The cornerstone is the *No-Surprise Rule*: any decision that changes the household's obligations, drains shared cash, or creates new debt is discussed *before* it happens. Not after. Not "I thought you'd be fine with it." Not "I didn't want to stress you out." Before.

This rule doesn't exist to control anyone. It exists to protect the relationship from the special kind of anger that shows up when you learn you've been financially committed without your consent.

Why money feels like a relationship issue (because it is)

Money fights are rarely about one transaction. They're about safety and power.

When one person is worried about bills, they're not just worried about bills—they're worried about the feeling of free-falling. When one person wants to spend, they're not just "being irresponsible"—they might be trying to claw back a sense of normalcy, dignity, or joy. If you miss that, every conversation becomes a courtroom drama: prosecution vs. defense. Who's right, who's wrong, who's the problem.

A better frame: money is a shared system under pressure. Pressure changes people. Under pressure, we interpret neutral events as threats. We catastrophize. We hide. We snap. This is why the household that waits to talk about money *until* it's urgent ends up talking about money in the worst possible conditions: late at night, after a long day, with one person already tense and the other already defensive.

A standing "money meeting" is not romantic. That's the point. It makes money boring again—administrative instead of existential.

The real enemy is ``financial fog''

Financial fog is what happens when nobody has the whole picture.

- One partner knows the credit card balance but doesn't mention it because it feels shameful.
- The other partner assumes savings is fine because nobody has said otherwise.
- Bills are on autopay, which is convenient, until the account overdrafts and everyone acts shocked.

- Each person has a private mental ledger of sacrifices they've made, and they're silently scoring the other person.

Fog turns small problems into big ones because it delays correction. It also turns couples into opponents: "Why didn't you tell me?" "Why didn't you notice?" "Why is this on me?"

Clarity is not just financial—it's emotional. When both people can see the same numbers, you can finally stop arguing about reality and start making choices.

The Monthly Money Date: 20 minutes, wine optional, panic not allowed

A money meeting is a short, regular check-in with rules. The most important rule is the tone: *no ambushes*. The second most important rule is the time limit. If you let it expand, it becomes a dreaded event, which means you'll avoid it, which means the problems will grow until the meeting becomes unavoidable and terrible. Keep it small on purpose.

Here's a simple cadence that works for many households:

- *Weekly* (*5 minutes*): "Anything weird coming up?" (quick scan of the calendar and known expenses)
- *Monthly* (*20 minutes*): the real meeting (review + decisions)
- *Quarterly* (*45 minutes*): bigger planning (insurance, subscriptions, upcoming trips, childcare changes, annual bills)

The monthly meeting is where the No-Surprise Rule lives. It's also where resentment goes to die—because instead of replaying the same argument every month ("we spend too much on food"), you write down a decision and treat it like a policy.

A concrete case study: how one couple stopped re-arguing the same points

Jordan and Elena had a recurring fight they could set a calendar to. It started the same way every time: one of them would notice the credit card balance had crept up. The other would respond with a perfectly true sentence: "We had a lot going on this month." Then the argument would turn into a messy pile of unrelated grievances: who does more for the kids, who is more stressed at work, who "gets" to buy coffee, who never wants to talk about money.

They weren't fighting about the credit card. They were fighting about *surprise* and *fairness*.

Two changes fixed it.

First, they created a 20-minute meeting every Sunday evening with an actual agenda. Second, they created a "decision log"—a shared note where they recorded agreements in plain language. Not paragraphs. One or two sentences. Example: "We keep groceries to $850 this month; if we exceed it, we pause dining out until next meeting." Another: "Any purchase over $200 from shared money requires a text first."

What changed wasn't their personalities. It was the number of times they had to renegotiate reality. The meeting created a neutral space for decisions. The decision log prevented their brains from turning every month into a fresh trial.

After about six weeks, the tone changed. The credit card balance stopped being a moral scorecard and became what it always was: a data point.

The No-Surprise Rule: what counts as a ``surprise"

You don't need to ask permission to buy toothpaste. The No-Surprise Rule is for things that change the household's financial gravity.

Common "surprise" categories:

- New debt: financing furniture, opening a card, "0% interest" offers, buy-now-pay-later plans
- New recurring commitments: subscriptions, gym memberships, kid activities with monthly fees
- Large one-time spending from shared money: appliances, travel bookings, gifts, car repairs (unless pre-agreed)
- Anything that changes fixed costs: moving, a new car payment, changing insurance tiers, childcare changes

The threshold is household-specific. Some couples use $100, some use $500. The right number is the one that prevents resentment. If you're arguing later, the threshold is too high. If you're constantly interrupting each other for tiny purchases, it's too low.

A useful approach is to set two thresholds:

- *Discuss first*: purchases over $X or new recurring payments
- *Inform within 24 hours*: anything unusual that could cause confusion ("I paid the registration today; it was $312.")

That second category is magic. It prevents the "What is *this* charge?" conversation, which is never fun and often accusatory even when nobody intends it that way.

The agenda that keeps meetings from turning into therapy

A money meeting fails when it becomes a referendum on the relationship. The goal is not to excavate your childhood in a 20-minute slot. The goal is to run the household like adults who like each other.

A tight agenda keeps it practical:

1) Start with wins (2 minutes). This sounds cheesy until you try it. Wins aren't bragging; they're proof you're not doomed. A win can be "we paid the card down $200," or "we didn't overdraft," or "we actually looked at the numbers." You're training your brain to associate money talks with progress instead of punishment.

2) Check the scoreboard numbers (6 minutes). Pick a handful of numbers you always review. For many households:

- checking balance (today)
- credit card balance(s)
- savings buffer balance
- any bill due in the next 10 days

This is not the time to categorize every expense. It's a quick health check.

3) Upcoming expenses (6 minutes). Look at the calendar. What's coming up in the next month that costs money? Birthdays, travel, school fees, car maintenance, annual subscriptions, co-pays. Write them down. Decide where the money will come from *before* it's due. This single habit eliminates a huge amount of "why is the card high again?" confusion.

4) Decisions (6 minutes). This is the heart of the meeting: choose one or two decisions and write them into the decision log. The most effective decisions tend to be structural, not emotional: adjust a transfer, pause a subscription, set a spending cap, pick the next debt to pay down, schedule a refinance call, assign who will handle an insurance quote.

If you leave the meeting with no decisions, you held a meeting to feel anxious together.

The decision log: your anti-amnesia device

Under stress, humans don't remember agreements well. We remember feelings. We remember that we felt criticized. We remember that we felt alone. We remember that we felt like the other person "promised" something. Two weeks later, you're arguing about whether the agreement ever existed.

The decision log is a shared, living record. It can be a note on your phone, a shared document, a whiteboard in the pantry—anything both people can access.

Good decision logs are:

- *Short*: one sentence per decision.
- *Specific*: "Transfer $150 each payday to the car repairs fund."
- *Time-bound*: "For February, dining out is $120 total."
- *Non-moralizing*: no "because you always..." language.

A decision log turns a relationship problem ("you never listen") into a systems problem ("we didn't write down what we decided"). Systems problems are solvable without shame.

Scripts that de-escalate conflict without swallowing your feelings

A good money system includes a few pre-written sentences—scripts you can borrow when emotions rise. The script isn't fake. It's a guardrail that prevents you from saying the thing you can't unsay.

Here are scripts that work because they name the problem without assigning villainy:

When you're surprised:

> "I'm feeling startled by that charge. Can you tell me what it was before my brain invents a story?"

When you're ashamed:

> "I don't feel great about this, but I want us to have the real numbers. Here's what happened."

When you need a pause:

> "I want to solve this, and I'm too heated to be useful. Can we take 20 minutes and come back?"

When values clash:

> "I think we're solving different problems. You're trying to buy relief; I'm trying to buy safety. How do we do a little of both?"

You're not trying to win. You're trying to stay on the same team.

Fairness: the hidden fuel of money conflict

Many money fights are actually labor fights in disguise.

If one person carries the mental load—tracking bills, remembering due dates, worrying about balances—they often feel like the household's unpaid CFO. The other person may feel controlled or treated like a child. Both positions are miserable.

A healthy money meeting includes explicit division of labor:

- Who checks the balances weekly?
- Who pays which bills (or monitors autopay)?
- Who calls the insurance company, negotiates the medical bill, schedules the car maintenance?

Rotate roles occasionally if one person is burning out. Not because everyone must do everything equally every week, but because shared competence creates shared power. Shared power reduces resentment, and resentment is a terrible financial advisor.

What to do when one partner hates talking about money

Avoidance is not laziness. It's often a stress response. Some people avoid money because they're afraid of conflict. Some avoid it because they feel behind and don't want to face the shame. Some avoid it because their childhood taught them money conversations are dangerous.

You can work with that instead of against it:

- Keep meetings short and consistent.
- Pick a calm time (not midnight, not right after work).
- Make the first goal "clarity," not "fix everything."
- Use shared visuals: one page, a few numbers, no interrogation.

A quiet but powerful move: agree that the meeting is not the place for sarcasm, eye-rolling, or character judgments. Those habits feel small in the moment and destroy trust over time. Trust is the only currency that makes joint finance work.

Actionable takeaway: schedule one meeting and install the No-Surprise Rule tonight

Before the day ends, do three things:

1. Put a 20-minute recurring money meeting on the calendar for the next four weeks. Same day, same time. Protect it like any other appointment.
2. Choose your No-Surprise threshold (start with something like $200) and define it in one sentence: *"Any pur-*

chase over $X from shared money, or any new recurring expense, gets discussed first."

3. Create a decision log in a shared note and write your first entry: the meeting time and the threshold.

That's enough to change the household atmosphere. Not because the numbers magically improve overnight, but because the dread starts to shrink. When surprises disappear, trust grows. When trust grows, you can talk about the real problems without turning each other into the problem.

2.4 Account Structure Strategy: Yours, Mine, and Ours

If you want to understand a couple's money system, don't ask for their budget. Ask a more revealing question: "Where does your paycheck land?"

That one detail quietly determines everything that happens next: whether bills get paid on time, whether one person becomes the household's unpaid accountant, whether spending turns into suspicion, and whether "we're fine" is based on actual visibility or on vibes.

Most households don't *design* an account system. They inherit one. They open a checking account when they're twenty-two. They add a spouse later. They tack on a savings account after an overdraft scare. They keep old accounts out of inertia. Then they wonder why money feels like a constant low-grade negotiation.

Account structure is not paperwork. It's relationship architecture.

The goal isn't to find the One True Way to bank. The goal is to build a system that does two things at the same time:

- *Transparency*: shared bills are visible and reliably covered.
- *Autonomy*: each partner has some spending freedom without a courtroom cross-examination.

That's why the best all-weather setup for many couples is a hybrid model: *Ours* for the household, plus *Yours* and *Mine* for personal discretion.

The three common account styles (and the fight each one invites)

Most couples land in one of these patterns:

1) Fully merged ("Everything in one pot").

All income flows into joint accounts; all expenses flow out.

This can be beautifully simple. It also demands a strong shared philosophy about spending, because every purchase is implicitly a statement about *us*. When it works, it feels like teamwork. When it doesn't, it feels like surveillance.

2) Fully separate ("Roommates with benefits").

Each person keeps their own accounts; they split bills through transfers or apps.

This can preserve independence, especially for couples who came in with established finances or different money personalities. The downside is stealth complexity: someone still has to track who paid what, what "fair" means, and what happens when life gets uneven (job loss, maternity leave, one partner going back to school). Separate systems often fail exactly when you need unity most.

3) Hybrid ("Ours for the machine, mine for the human").

Income funds a joint hub that pays household obligations; personal accounts receive a predictable transfer ("allowance") for guilt-free spending.

Hybrid models reduce the two most common sources of conflict:

- *Fear*: "Are we going to cover the bills?"
- *Control*: "Do I have to justify every nonessential purchase?"

Hybrid isn't about secrecy. It's about removing friction from the relationship while keeping the shared system visible.

Why ``separate" often turns into ``secret," even when nobody means it to

People don't hide spending because they're evil. They hide spending because they're tired.

If every purchase comes with commentary—*"Do we really need that?" "How much was it?" "You spent again?"*—eventually a normal adult will start routing around the criticism. Not because they want to betray their partner, but because they want to buy a pair of shoes without a debate.

A hybrid model makes room for harmless individuality. It creates a safe zone where a latte is just a latte, not a referendum on discipline.

The key phrase is *no-questions-asked money*. That doesn't mean "I never tell you anything." It means: once the household obligations are funded, each person gets a defined amount that they can spend without needing approval or defending their taste.

Ironically, this often increases honesty. When people don't feel policed, they volunteer information more freely.

The non-negotiable safety feature: contingency access

Every household should plan for the boring disasters:

- A partner is hospitalized.
- A partner is traveling when the mortgage payment fails.
- A partner dies.
- A partner is dealing with depression and can't handle paperwork for a while.

A system that requires one person to be constantly functional is not a system. It's a single point of failure.

Whatever structure you choose, build in contingency access:

- Both partners can log in to the joint hub account.
- Both partners know where the bill list lives.
- Both partners can access an emergency fund without pleading for credentials.

Independence is nice. Redundancy is love.

The Hybrid Account Model (the blueprint)

Here is the cleanest version of the hybrid architecture:

- *Income lands in the Joint account* (the hub).
- *Bills are paid from the Joint account* (housing, insurance, childcare, minimum debt payments, utilities-your household "machine").
- *Two predictable transfers go out to Personal accounts* (one for each partner).

The transfers are the heartbeat. They convert "we should be careful" into an automatic boundary. If a personal account runs low, that person has a decision to make-without destabilizing the household bills.

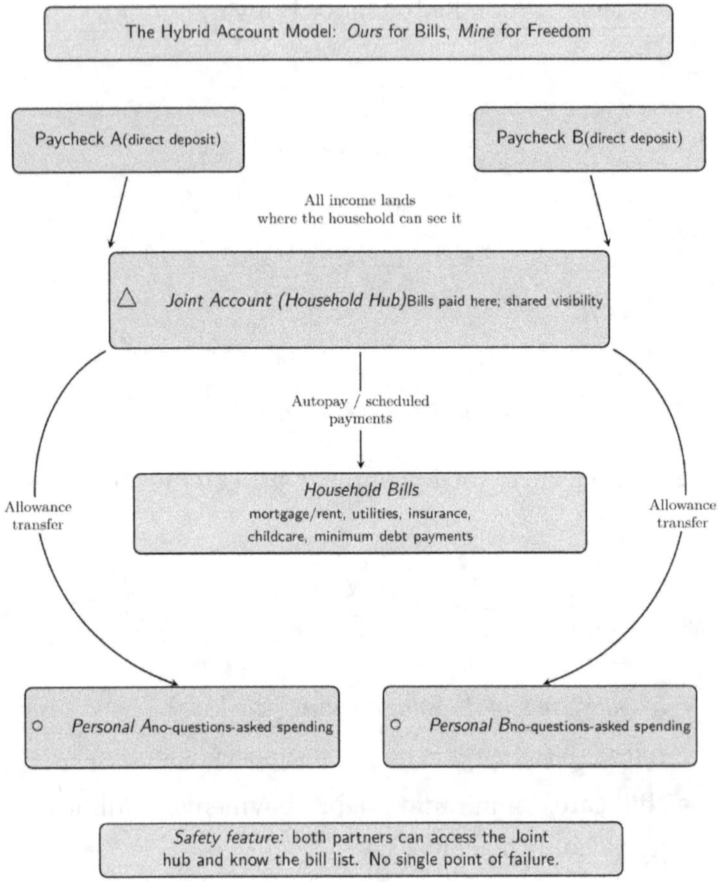

Notice what's *not* in this diagram: morality. There's no "good spender" and "bad spender." There's just flow. The system does the hard work of preventing a few common disasters:

- Household bills don't depend on one person remembering due dates.
- Personal purchases don't require a mini-approval process.

- The shared picture stays shared, reducing financial fog.

Case study: the hybrid model that saved a marriage from a thousand paper cuts

Sam and Priya didn't have a huge money problem. They had a *constant* money problem.

Sam paid most bills from his checking account because he was "better with numbers." Priya Venmoed her half when she remembered, which made Sam feel like he was managing the household alone. Priya felt treated like a teenager being monitored. They were both technically right, which is the most exhausting kind of conflict.

They tried a shared spreadsheet. It lasted three weeks. They tried splitting expenses more "fairly." It turned into a recurring debate about what counted as "household." Was the dog's vet bill household? What about Priya's work clothes? Sam's lunches? Every month, another little negotiation.

Their breakthrough was structural, not emotional. They set up:

- a joint checking account for bills
- a joint savings account as a buffer
- two personal checking accounts

Both paychecks were redirected so a fixed amount landed in the joint hub each payday (enough to cover the monthly nut plus a buffer). Then each personal account got the same "allowance" transfer every payday.

Two months later, they reported something that surprised them: they fought *less* about money, even though their total spending hadn't dramatically changed yet. The system reduced the number of moments where they had to interpret each other's choices. It also changed Sam's role from "bill cop" to "system maintainer," which felt completely different.

The hybrid model didn't fix their relationship. It stopped their banking setup from *damaging* it.

How big should the personal allowance be?

Allowances sound juvenile until you experience how adult it feels to have spending freedom without tension.

There's no universal number. What matters is that it's:

- *predictable* (same schedule, same amount)
- *fair* (equal dollar amount is simplest; proportional can work but needs goodwill and clarity)
- *funded after essentials* (bills first, then freedom)

If money is tight, the allowance can be small. Even $25/week per person can reduce friction because it creates a bright line between "shared obligations" and "personal choices." If money is less tight, the allowance can support hobbies, lunches, gifts, or the harmless personal spending that makes you feel like a person instead of a household expense manager.

One rule keeps the allowance from turning into a loophole: big purchases that create future obligations still trigger the No-Surprise Rule. Buying concert tickets from your personal account is fine. Financing a new phone on a payment plan that will hit the household cash flow next month is not a personal-only decision.

What about uneven incomes?

Uneven incomes are normal. The trap is treating uneven income as uneven *voice*.

A hybrid model can respect uneven incomes without turning the higher earner into the "owner" of the household. Common approaches:

- Fund the joint hub proportionally to income, but keep equal personal allowances.
- Fund the joint hub to cover the monthly nut and shared goals, then decide together how much remains for additional saving or lifestyle upgrades.

The emotional north star: both partners should feel like full adults with agency, not like an employee requesting funds. If your system makes one person feel small, it will eventually cost you somewhere else.

Practical setup details that prevent avoidable chaos

A good structure is only as good as its execution. A few tactical choices make the hybrid model smoother:

Use the Joint hub as the autopay center.

If bills come out of one place, you eliminate the "Who paid the electric bill?" confusion. Autopay isn't mindless if you have visibility; it's just efficient.

Keep the Joint account boring.

The Joint hub is not the place for impulse purchases. The more boring it is, the more reliable it becomes. If you want the household to feel stable, treat the joint account like the engine room, not the snack bar.

Build a small buffer into the Joint hub.

A joint account that runs near zero is a stress factory. Even a modest cushion changes everything: fewer overdrafts, fewer "timing" crises, fewer late fees.

Name your accounts like a grown-up.

"Checking 1" is how you end up transferring money to the wrong place at 11:47 p.m. Name accounts by purpose: "Bills Hub," "Emergency Buffer," "Car Repairs," "Priya Fun," "Sam Fun." Your future self will thank you.

Make sure both partners can access everything they must access.

Passwords, two-factor authentication, where the bill list lives-these aren't romantic topics, which is why they're so easy to skip. Then life forces the issue at the worst time.

When hybrid doesn't work (and what to do instead)

Hybrid isn't magic. It struggles in a few scenarios:

Active addiction or compulsive spending.

"No-questions-asked money" is not appropriate if spending is out of control and harming the household. In that situation, you need guardrails, accountability, and often professional support. Structure should reduce friction, not enable destruction.

Ongoing secrecy or financial infidelity.

If someone is hiding debt or accounts, the problem is not account design; it's trust. The system can support rebuilding trust, but it can't substitute for it.

Extremely tight cash flow.

When every dollar is already committed, allowances may feel impossible. In that case, the "allowance" might be tiny at first, but the structure still helps because it makes the shortage visible and shared rather than silently carried by one partner.

Even then, the hybrid idea can be scaled down: one joint bills account plus personal accounts, with small transfers and strict clarity. The point is the flow, not the size.

Actionable takeaway: a 45-minute account redesign that reduces friction immediately

Set aside 45 minutes and make three decisions on purpose:

1. Choose your household hub: one Joint checking account that will pay all shared bills. Write down which bills count as "shared" (start with housing, utilities, insurance, childcare, minimum debt payments, groceries).
2. Choose your autonomy amount: set a personal allowance (even small) and decide the transfer schedule (each payday is easiest).
3. Install the safety feature: ensure both partners can log into the Joint hub, know where the bill list lives, and can access emergency cash.

Then do one concrete action within the next 24 hours: switch one major bill to the Joint hub, or redirect one paycheck to land there, or set up the first automatic allowance transfer. A good account structure doesn't just organize money—it lowers the daily temperature of your household, so financial decisions stop feeling like personal critiques and start feeling like what they are: choices you make together.

2.5 Automation and Buffers: The Self-Driving Wallet

A lot of financial stress isn't caused by *how much* you spend. It's caused by *when* life demands the money.

Your car insurance is cheap—until it arrives as a six-month premium that lands in the same week as a birthday party, a dentist appointment, and a mysteriously urgent school fundraiser. Your holidays are "manageable"—until November shows up like a freight train and your checking account becomes a crime scene. Even "routine" home maintenance has a talent for clustering: the fridge makes a noise, the dryer stops drying, and the toilet runs like it's training for a marathon.

Households don't usually spiral because they can't afford life in the long run. They spiral because life sends invoices in bursts. And bursts create two enemies: surprise and decision fatigue.

Decision fatigue is the quiet monster. It's the exhaustion that comes from answering the same question over and over: *Can we afford this right now?* When you're tired, you don't make calm tradeoffs. You make fast ones. You swipe a card "just this once." You skip the transfer to savings and promise you'll "catch up." You pay the late fee because you forgot the due date again. None of these choices feels like a turning point—until you add them up.

Automation and buffers are how you stop running your household like an emergency room.

A *self-driving wallet* is a system where the right money moves happen by default: bills get paid on time, savings grows quietly, and irregular expenses are pre-funded so they arrive as boring transactions instead of mini-crises. The goal isn't perfection. The goal is to make your household resilient on your worst Tuesday, not your best Sunday.

Why willpower is a terrible financial plan

The classic approach to money management is essentially: *try harder*. Check the account more often. Be more disciplined. Communicate better. Make fewer "bad" choices.

That approach fails for the same reason crash diets fail. Humans are not machines. You will have months where you're sick, stressed, slammed at work, grieving, sleep-deprived, or simply burned out. Those are not moral failures. They're the standard operating conditions of adult life.

If your household only "works" when you are vigilant, rested, and hyper-attentive, it doesn't work.

Automation is what you build when you accept a realistic truth: you can't out-discipline randomness forever. But you *can* out-system it.

Buffers: the money equivalent of shock absorbers

A buffer is cash set aside for a specific purpose so you don't have to invent money at the last minute. Buffers turn "unexpected" expenses into expected ones. They don't prevent the pothole; they keep it from blowing out your tire.

There are two main buffer types:

1) **Emergency buffer.** This is the "something went wrong" fund: job loss, medical issue, car repair, urgent travel. It's not for predictable expenses; it's for the ones that change your month without asking permission.

2) **Sinking funds.** A sinking fund is money you set aside gradually for a predictable-but-irregular expense. The expense is not an emergency; it's an inevitability that refuses to show up monthly.

Sinking funds are the secret sauce because they solve the most common household problem: "We're fine most months, but then we get hammered."

The idea that changes everything: bill smoothing

Bill smoothing is simple: instead of paying a $600 bill when it arrives, you save $100/month for six months and pay it from the buffer. You transform a spike into a flat line.

What changes isn't just the math. It's your nervous system.

Spending spikes create shame ("Why can't we handle normal life?") and conflict ("You should have planned for this"). A smoothed system creates calm: the bill arrives, you pay it, nothing dramatic happens. You don't need a family summit to decide whether to pull from savings or swipe a card. Past you already decided.

A household with bill smoothing feels wealthier than its income would suggest because it experiences fewer financial shocks.

A concrete example: turning the ``rollercoaster" into a predictable month

Consider a household that regularly gets hit with these irregular expenses:

- Car insurance: $720 every 6 months
- Vehicle maintenance/repairs (average): $900 per year
- Gifts and holidays: $1,200 per year
- Annual memberships/subscriptions: $240 per year
- Back-to-school costs: $400 per year

Total: $3,460/year, or about $288/month.

Without sinking funds, that $3,460 shows up as sudden spikes: a brutal month here, a bruised month there, and a lot of "we'll figure it out." With sinking funds, you transfer $288/month into labeled buckets and pay those expenses like clockwork.

The household didn't magically become richer. It became less surprised.

Case study: the couple who stopped using credit cards as a bridge

Nina and Marcus were the kind of people who said, sincerely, "We don't spend that much." They weren't lying. Their day-to-day spending was reasonable. The problem was that their credit cards functioned like a temporary dam: whenever an irregular expense hit, they put it on the card and planned to pay it off "over the next couple of paychecks."

In theory, that worked. In practice, the next irregular expense always arrived first.

The worst month was not catastrophic. It was ordinary: a $480 car repair, a $210 annual subscription renewal they forgot about, and a $300 plane ticket for a family event. The card balance jumped, and then interest started charging them for the privilege of having a normal life.

They switched to a self-driving wallet in three moves:

1. They created a single "Bills + Buffers" joint account that received automated transfers each payday.
2. They built three sinking funds first (car, gifts/holidays, annual bills) and ignored everything else until those were stable.
3. They left their credit cards for convenience but stopped treating them as funding sources. Every swipe was backed by money already sitting in the buffer account.

The surprising benefit was emotional: Marcus stopped feeling like Nina was "always spending," and Nina stopped feeling like Marcus was "always anxious." The system handled what their personalities couldn't: consistent preparation.

Three months later, they weren't debt-free. But they were no longer adding new debt for predictable expenses. That's the moment the future starts to change.

Where automation belongs (and where it doesn't)

Automation is powerful, but it's not a substitute for awareness. Think of it like a thermostat: you set it, it runs, and you still glance at it occasionally to make sure it's doing what you intended.

Good candidates for automation:

- Fixed bills (mortgage/rent, insurance, utilities)
- Minimum debt payments (so you never accidentally miss one)
- Sinking fund transfers (small, steady, boring)
- Emergency fund contributions (even if modest)

Poor candidates for automation (at least early on):

- "Whatever is left" investing transfers that might cause overdrafts
- Aggressive debt extra-payments before cash flow is stable
- Autopay on credit cards if you don't have a buffer and tend to run tight (it can create timing chaos)

A useful principle: automate what you *must* do, and what you will reliably be glad you did. Don't automate aspirational transfers that create emergencies. A system that bounces payments is not "disciplined"; it's brittle.

The simplest self-driving wallet design

You don't need twelve accounts and color-coded spreadsheets. You need a clear flow.

A practical setup many households can maintain:

- *Bills account* (*joint*): receives enough to cover fixed costs + planned variable essentials.
- *Buffer/sinking funds account* (*joint savings*): holds emergency fund and sinking funds (can be one account with labeled "buckets," or separate savings accounts).
- *Spending accounts* (*personal*): each partner gets a predictable transfer for no-questions-asked spending.

Then schedule the moves:

- Paychecks land (or immediately transfer) into the Bills account.
- Bills are paid on autopay from Bills.
- A fixed amount transfers to Buffers shortly after payday.
- Personal allowances transfer on payday (or the day after).

If you already have a hybrid account structure, automation is the layer that makes it feel effortless instead of fragile.

Building sinking funds without getting overwhelmed

The fastest way to quit sinking funds is to create too many of them. Start with the ones that have historically punched you in the face.

A high-impact starter set:

- *Car repairs/maintenance*
- *Holidays and gifts*
- *Annual/semiannual bills* (insurance premiums, subscriptions, professional dues)

Once those are running smoothly, you can add:

- *Medical out-of-pocket*
- *Home maintenance*
- *Travel*
- *Back-to-school*

To keep it grounded, build each sinking fund using the same formula:

$$\text{Monthly contribution} = \frac{\text{Expected annual cost}}{12}$$

The psychological trick: move money when you're least likely to miss it

The best time to automate transfers is right after payday—before money gets mentally assigned to spontaneous spending.

If you wait until the end of the month to "save what's left," you're asking leftover money to exist in a world where life is actively trying to eat it.

Payday-based automation has another advantage: it matches human behavior. Most people don't decide to "spend too much" at once. They make a hundred small choices that feel harmless because the balance looked fine *at the time*. Moving money early creates a meaningful constraint without constant self-control.

How to tell if your buffers are working

A buffer is doing its job when:

- You can pay irregular bills without using credit cards.
- Your checking account doesn't swing wildly between "fine" and "terrifying."
- Money meetings focus on choices, not damage control.
- You stop treating every surprise as evidence you're failing.

The metric isn't perfection. It's fewer emergencies created by predictability.

Actionable takeaway: set up one transfer that eliminates one recurring crisis

Pick *one* expense that has repeatedly messed with your month—car insurance, holidays, annual subscriptions, back-to-school—and do the following today:

1. Estimate the annual cost (use last year's number, or a conservative guess).
2. Divide by 12 to get a monthly transfer.
3. Create a labeled sinking fund bucket (in a savings account or a dedicated sub-account).
4. Automate the transfer to happen within 24 to 48 hours after each payday.

Then make a simple rule: when that bill arrives, you pay it from the sinking fund, not from hope. The first time you do this successfully, you'll feel something rare in adult finance: the quiet satisfaction of a plan that worked exactly as intended.

Chapter 3

Escaping the Fixed-Cost Trap

It is a common paradox: earning two salaries but feeling more financially precarious than your parents did on one. The culprit isn't usually too many lattes; it is the 'Fixed-Cost Trap.' When we lock our future income into rigid monthly bills-outsized mortgages, two car payments, childcare, and subscriptions-we lose the flexibility to navigate life's bumps. This chapter moves beyond standard budgeting advice to focus on 'lowering your nut.' We will dismantle the biggest expenses first, transforming your household from a fragile machine that requires perfect conditions to run, into a resilient engine capable of weathering job losses, health scares, or career pivots.

3.1 Minimizing Fixed Costs

A funny thing happens when you get a raise: your life gets easier for about three months.

Then the new normal shows up. The nicer apartment that "just makes sense." The car with the payment that "fits the budget." The subscription bundle that "saves money," because it replaces a few other things. The upgraded phone plan, the premium gym, the pricier daycare that feels safer, the new insurance policy with the better coverage. None of these choices are obviously reckless. Most of them are completely understandable. They also have one dangerous thing in common: they move from *optional* to *owed*.

That's the heart of the fixed-cost trap. It's not that you spend money. It's that you *commit* money.

The ``Monthly Nut": the bill pile that owns you

Old-school sailors used "the nut" to mean the basic cost of running the ship-what had to be paid before anyone got paid. Households have the same thing. Your monthly nut is the amount of money required to keep your life from falling apart: housing, basic utilities, minimum debt payments, insurance, childcare, transportation that gets you to work, and anything else that has real consequences if you miss it.

The nut isn't the same as "your expenses." Groceries can be adjusted. Entertainment can be trimmed. A vacation can be postponed. But the nut has teeth. Miss the rent or mortgage and you don't just "fall behind." You invite late fees, credit damage, collection calls, eviction or foreclosure risk, and a stress level that makes every other decision worse. Fixed costs don't merely take money-they take *freedom*.

There's a particular kind of anxiety that comes from knowing your next paycheck already has a job: it's not "paying you," it's feeding a machine you built on purpose.

Commitments vs. choices: why the category matters more than the dollar

A fixed cost is not "a bill that's the same every month." It's a commitment that's hard to change quickly without pain. Some are truly fixed (a mortgage payment on a fixed-rate loan). Some are semi-fixed (car insurance you can shop, but not instantly). Some are disguised fixed costs wearing a hoodie (a "0% financing" deal that quietly locks you into a payment schedule).

Variable costs are choices. Not always pleasant choices, but choices you can make this week without calling a lawyer. If your grocery bill is too high, you can change what you buy. If your housing cost is too high, changing it takes time, money, and emotional energy-and sometimes it's simply not possible on short notice.

This is why two households with the same income can live in completely different financial universes. One is driving with a flexible suspension. The other is driving a sports car with bald tires: fast on a smooth road, terrifying on a pothole.

The fragility ratio: how close are you to ``one bad month"?

Here's a simple way to measure how precarious your situation is:

$$\textit{Fragility Ratio} = \frac{\text{Fixed Costs}}{\text{Take-Home Pay}}$$

Take-home pay is what actually lands in your checking account after taxes and payroll deductions. Fixed costs are the bills you can't quickly reduce without serious fallout.

This ratio does something a budget app can't: it tells you how much of your life is already spoken for. If your fragility ratio is 0.35, you have room to maneuver. If it's 0.70, you are living inside a narrow hallway: one layoff, one health problem, one childcare disruption, one car repair-pick your villain-and you're suddenly negotiating with creditors.

There's a cruel twist here. High fixed costs don't just create financial risk; they create decision risk. When you have no margin, you make choices under pressure. You take the "quick fix" debt. You skip the preventive doctor visit. You ignore the weird noise in the car. You don't leave a job that's grinding you down because the bills don't care how you feel. A high fragility ratio doesn't just predict money problems; it predicts a life that's harder to steer.

A concrete case: two households, one income drop, two different futures

Consider Maya and Jordan, and Sam and Riley. Both households bring home $7,500 per month. Both live in the same metro area. Both feel "middle class." Both are good people who love their kids and sometimes order takeout because they're exhausted.

The difference is what they promised their future paychecks.

Maya and Jordan's fixed costs:

- Mortgage + property tax + insurance (escrowed): $3,200
- Two car payments: $900
- Auto insurance: $280

- Childcare: $1,400
- Student loan minimums: $450
- Phones + internet: $220
- Basic utilities (hard to change quickly): $300

That's $6,750 in mostly non-negotiable monthly commitments. Their fragility ratio is 0.90.

Sam and Riley's fixed costs:

- Rent: $2,200
- One car payment + one older paid-off car (maintenance sinking fund treated as fixed): $450
- Auto insurance: $220
- Childcare: $1,200
- Student loan minimums: $350
- Phones + internet: $160
- Basic utilities: $280

That's $4,860. Their fragility ratio is 0.65.

Now the shock: a 20% income drop. Maybe a layoff, maybe commissions dry up, maybe hours get cut, maybe a parent needs time off. Take-home pay falls from $7,500 to $6,000.

Maya and Jordan are instantly underwater. Their fixed costs ($6,750) exceed their income ($6,000) by $750 before groceries, gas, prescriptions, school supplies, or anything resembling a normal human life. They don't have a "tight month." They have a math problem. And math problems don't respond to optimism.

Sam and Riley feel the hit too. But their fixed costs ($4,860) still leave $1,140 for the messy middle: food, fuel, and the unpredictable stuff. It's not comfortable, but it's survivable without immediately sliding into high-interest debt. They can trim variable spending, pause extra savings, renegotiate a bill or two, and buy time.

Same shock. Different outcome. Not because one couple is more disciplined, but because one couple is less committed.

Why cutting fixed costs is 10x more powerful than ``being good"

A lot of personal finance advice quietly assumes your problem is *behavior*. Spend less on lattes. Pack lunches. Cancel the streaming service you barely use. Those are fine moves, and sometimes they matter. But they're also the financial equivalent of trying to bail out a boat with a coffee cup while the leak is still open.

Fixed costs are the leak.

When you reduce a fixed cost, you don't just save money once. You lower the minimum required to breathe every month. That improvement compounds. A $300 lower housing payment doesn't just create $300 of savings. It creates:

- more resilience in emergencies,
- more capacity to pay down expensive debt,
- more ability to handle a job change without panic,
- and more room to save for the future without feeling punished.

This is why fixed-cost reduction can feel like a cheat code. It's not glamorous. It rarely gets applause. But it changes the physics of your household.

Also, fixed costs are where "lifestyle inflation" hides. People don't usually inflate their lifestyle by buying twenty designer jackets. They inflate it by upgrading the monthly obligations: better zip code, newer car, more expensive insurance, bigger house, premium everything. The spending looks sensible because it comes with reasons. And then one day the reasons don't matter, because the obligation remains.

The sneakiest fixed costs: the ones you call ``normal"

Some fixed costs announce themselves loudly: rent, mortgage, car payment, childcare. Others sneak in through the side door.

The subscription stack. A few $9.99 charges feel harmless until you realize you're funding an entire second cable bill. The bigger issue isn't the total; it's the automatic nature. Autopay turns a decision into a default. Defaults run your life when you're tired.

The "monthly" version of everything. Furniture, phones, mattresses, tires, lawn equipment-modern retail is built to turn purchases into payments. A payment plan is a fixed cost in disguise. It also encourages you to buy more than you would if you had to feel the full price in one moment.

Insurance creep. Some coverage is essential. Some is peace-of-mind shopping when you're anxious and busy. If you've ever said, "It's only $18 more per month," you've felt the trap: the human brain is terrible at adding up "only" across twelve months and ten categories.

Minimum payments. Debt minimums are fixed costs that you gave your past self permission to assign to your future self. If your life is too tight to pay extra, the minimum becomes permanent, and "temporary" debt turns into a long-term tenant.

The second car. It's rarely just the payment. It's insurance, registration, maintenance, repairs, tires, and the slow leak of convenience spending that comes with driving more. Two incomes plus two cars is one of the most common "we're doing well" symbols that quietly makes families fragile.

Fixed costs and two-income households: the special kind of risk

Two incomes feel like a safety net. In one sense, they are. In another, they can be a trap with better lighting.

When both incomes are necessary to cover the nut, the household becomes a high-wire act. If one person loses a job, or needs time off, or gets sick, or a family member needs care, the entire structure shakes. The budget isn't "shared." It's *interlocked.* That interlock creates pressure: the higher-earning spouse may feel trapped in a job; the lower-earning spouse may feel guilty for any interruption; both may feel like they can't breathe.

There's also the childcare boomerang. Two incomes can require paid care. Paid care becomes a major fixed cost. And then the household can't reduce work hours because the nut is too high. You end up working to afford the fixed cost that lets you work. That loop isn't anyone's fault. But it's a loop you want to see clearly, because you can't exit a maze you refuse to name.

A practical way to find your fixed-cost ``kill switch"

If you only do one thing after reading this, do this:

Write down the number that would ruin your month.

Not your total spending. Not your ideal budget. The number that, if your income fell below it, would trigger missed payments, borrowing, or a slow-motion panic.

To find it, list these categories and add them up:

- Housing payment (rent or mortgage) + required insurance/taxes/HOA
- Minimum debt payments (credit cards, student loans, personal loans)
- Transportation required to work (car payment, insurance, transit pass)
- Childcare required to work
- Essential utilities (electric, gas, water, basic internet/-phone)
- Any contract you can't realistically cancel in 30 days

That total is your nut. Now subtract it from your take-home pay. The leftover is your margin.

Margin is not "extra." Margin is the difference between an annoying problem and a crisis.

If your margin is small, you don't need a pep talk. You need a structural change.

Where to cut first: the ``big three" that move the needle

Not every fixed cost is equally cuttable. The trick is to focus on the categories where a change actually matters.

Housing. This is usually the biggest lever, and it's also the most emotional. Cutting housing costs can feel like moving backward, even when it's moving you toward stability. But a household with affordable housing can survive almost anything. A household that's house-poor can be taken down by a single surprise.

Transportation. Car payments are a fixed-cost factory. A reliable older car can be a financial superpower, even if it's not the one you'd pick on a showroom floor. If you have two car payments, you have two anchors.

Debt minimums. Minimum payments are fixed costs you can kill. If you're carrying high-interest debt, the goal isn't "a balanced budget." The goal is to reduce the nut by eliminating mandatory payments. Every paid-off loan is a future pay raise.

Notice what's not on the list: the small stuff. You can absolutely trim it, and it can help. But if you're trying to reclaim your life, start with the levers that actually lift weight.

The emotional hurdle: why fixed-cost cuts feel like identity loss

People don't cling to fixed costs because they love bills. They cling to what the bills represent: safety, status, convenience, a good childhood, a sense of progress, a neighborhood you're proud of, a car that starts every time, a home where guests can stay, a room for each kid.

When someone suggests downsizing, what you hear might be: "Give up the life you worked for."

So it helps to rename what you're really buying.

A bigger house is often a purchase of *certainty*: certainty that you won't have to move again, certainty that you'll have space, certainty that you're doing what responsible adults do. A new car is often a purchase of *control*: control over breakdowns, control over embarrassment, control over whether you can reliably get to work.

Those needs are real. The twist is that high fixed costs can sabotage the very feeling you're trying to buy. When your nut is huge, your life becomes less safe, less controlled, and less certain. The bills you chose for comfort become a source of chronic stress.

Stability isn't the nicest version of your life. Stability is the version you can keep, even when things get weird.

A simple ``fixed-cost audit" that doesn't require becoming a new person

You don't need a spreadsheet masterpiece. You need a short, honest review.

Take your last two months of bank and credit card statements and highlight every recurring payment. Then sort them into three buckets:

- *Hard fixed:* must be paid; hard to change quickly (rent/-mortgage, childcare contract, debt minimums).
- *Soft fixed:* recurring, but you could change within 30–60 days if you had to (insurance, phone plan, internet, some subscriptions).
- *Disguised fixed:* not technically recurring, but predictable and inevitable (car repairs, medical deductibles, annual fees, school costs).

That third bucket is where many budgets go to die. If you ignore predictable irregular costs, you end up "surprised" every year by the same surprises. Treat them like mini fixed costs by saving monthly into a sinking fund (a boring name for a powerful idea: money set aside in advance for known future expenses).

Then ask two questions that cut through the noise:

1. *If our income dropped 20% next month, what would we be forced to do?*
2. *Can we make any of those changes now-calmly, on our terms-instead of later in a panic?*

The goal isn't austerity. It's optionality.

The ``danger zone'' test: can your household survive a normal shock?

A normal shock is not a meteor. It's the stuff that happens all the time:

- one income pauses for three months,

- commissions slump,
- a job search takes longer than expected,
- a parent needs support,
- a medical issue changes your schedule,
- the car needs a $1,800 repair.

If a normal shock would push your income below your fixed costs, you're in the danger zone. That doesn't mean you've failed. It means your system is fragile.

And fragility is fixable, but it usually requires one of three moves:

- lower the nut,
- increase stable income,
- build liquidity so you can absorb shocks without debt.

Households often try to skip the first move because it's uncomfortable. They go straight to hustling more hours. That can work temporarily. But it also burns people out, and it keeps the underlying structure intact. A household can't out-earn a badly designed cost structure forever, because life eventually demands time.

Actionable takeaway: design a ``shock-proof" number and protect it

Pick a number: the maximum fragility ratio you're willing to live with for the next year.

For many households, 0.60 is a powerful target. That means fixed costs are no more than 60% of take-home pay. You may not be there now. The point is to choose a direction that favors resilience over appearances.

This week, do three things in one sitting:

1. Calculate your take-home pay (average of the last 3 months).
2. Calculate your monthly nut (the commitments that bite if missed).
3. Choose one fixed-cost change that reduces the nut within the next 60 days-shop insurance, downgrade a phone plan, refinance a high-interest loan if possible, sell a payment car for a paid-off car, renegotiate a childcare arrangement, take in a roommate, move when the lease ends, anything that lowers the number you must earn to stay afloat.

You're not trying to win a contest for "lowest spending." You're trying to buy back the most valuable asset a household can have: the ability to handle a bad month without it turning into a bad year.

3.2 The Housing Decision: Rent vs. Buy

The most expensive sentence in American personal finance might be: *"Renting is throwing money away."*

It sounds wise because it contains a truth-shaped object: when you rent, you don't build home equity. But it leaves out the rest of the story—the part where homeownership has plenty of money you "throw away," too. The difference is that homeowners often throw it away quietly, bundled inside a payment they call *"investing."*

A better question than *"rent or buy?"* is: *"Which choice buys you more stability for your life right now, at a cost you can truly carry?"* The point isn't to win a cultural argument. The point is to avoid becoming the proud owner of a house that owns you.

Housing is a place to live *and* a risk you agree to carry

Most money decisions are reversible. Housing is not. It's a large, long-term bet that shapes your monthly obligations, your flexibility, and your ability to handle a bad surprise without turning it into a financial emergency.

Buying a home can be wonderful. It can also be a high-stakes concentration of risk: one asset, one location, one set of property taxes and insurance premiums you don't control, one roof that will eventually demand a five-figure tribute. Renting, by contrast, is often treated like a moral failing when it's really a contract that buys you a certain kind of freedom: the freedom to move, the freedom to cap your downside, and the freedom to keep your "monthly nut" from becoming a choke collar.

If you're deciding between renting and buying, you're not just choosing a building. You're choosing a financial *operating system.*

The myth hides in a missing comparison: rent vs. the *unrecoverable* costs of owning

Here's the simplest way to puncture the "throwing money away" myth:

When you rent, the entire rent payment is unrecoverable. When you own, a large portion of the monthly cost is also unrecoverable—just better disguised.

Unrecoverable costs are expenses that do not come back to you as equity. They are the price of using the home, not owning it. The big ones:

- *Mortgage interest* (especially early on, when the loan balance is high)
- *Property taxes*
- *Homeowners insurance*
- *Mortgage insurance* (PMI) if your down payment is small
- *Maintenance and repairs* (the "nothing stays fixed forever" fund)
- *HOA dues* (in many neighborhoods, unavoidable)
- *Closing costs* when you buy, and agent commissions/-transaction costs when you sell

If you buy a home and your monthly payment is $3,000, it is not accurate to say "I'm investing $3,000." Some of that is principal (which becomes equity). Much of it is the price of having a roof you control. That "price" can be entirely reasonable. It just needs to be named honestly.

Rent isn't throwing money away. Rent is paying for shelter with an explicit price tag and a built-in escape hatch.

A homeowner's payment is not one payment---it's a stack of payments

Many first-time buyers fall in love with the mortgage principal-and-interest number because it looks clean. But real housing cost is messier. A useful mental model is the full monthly ownership stack:

$$\begin{aligned} PITIA &= \text{Principal} + \text{Interest} + \text{Taxes} + \text{Insurance} \\ &+ \text{Association dues} \end{aligned}$$

That acronym isn't meant to be cute. It's meant to prevent the most common homebuying mistake: treating the mortgage payment as the whole cost and then acting surprised when escrow, taxes, and insurance show up like uninvited guests who refuse to leave.

Even within PITIA, there's another invisible line item: maintenance. A house has a way of turning "optional projects" into mandatory ones. You can postpone cosmetic upgrades. You can't negotiate with a leaking water heater.

If you're choosing between rent and buy, compare rent to *PITIA plus maintenance,* not to the mortgage payment you saw in an online calculator.

A concrete case: ``We can afford it" until the house starts sending invoices

Here's a real-world pattern, with numbers that look like a win until they don't.

A couple—call them Elena and Marcus—rent a two-bedroom for $2,400. They're tired of the annual rent increase email. They find a starter home listed at $420,000. They scrape together a 5% down payment. The lender quotes a monthly principal-and-interest payment around $2,500. That number is dangerously seductive because it feels like a straight swap: "We're already paying $2,400. For $100 more, we build equity."

But then the ownership stack arrives:

- Principal + interest: $2,500
- Property taxes (varies widely): $450
- Homeowners insurance: $180
- PMI: $220
- HOA: $0 (they got lucky)

PITIA: $3,350.

Add maintenance. A conservative planning number is 1% of home value per year (sometimes less, sometimes much more), which would be $4,200/year or $350/month. Now you're at $3,700.

So the true comparison wasn't $2,400 rent versus $2,500 mortgage. It was $2,400 rent versus something closer to $3,700 in ownership cost—before a single paint color is chosen.

Elena and Marcus might still decide to buy. But if they buy while believing the $2,500 number, they aren't purchasing a home. They're purchasing a budget crisis with granite countertops.

The rent vs. buy decision is really a time-horizon decision

Transaction costs are the silent killers of "buying as an investment." Buying and selling a home isn't like buying and selling an index fund. It's slow, expensive, and subject to the mood swings of local markets.

If you may move in a few years—career uncertainty, family plans, a possible return to school, caring for parents, a relationship that's still becoming itself—renting often functions like an insurance policy. You pay for flexibility. Flexibility is not a luxury. It's a risk-management tool.

Buying tends to work better when your time horizon is long enough to spread those transaction costs across many years, and when your life is stable enough that the house won't turn into a forced sale.

A forced sale is the nightmare scenario: you have to sell because of a job change, divorce, illness, or financial strain. Forced sales tend to happen at the exact moment you least want them to happen—when you're stressed, when the market may be down, and when you have no spare cash to make the home "market-ready." Renting can't protect you from every hardship, but it can reduce the odds that housing becomes the amplifier.

Buying can build wealth---but it's not a guaranteed wealth machine

Homeownership has two wealth-building engines:

1. *Forced savings*: you pay down principal over time.

2. *Price appreciation*: the house may be worth more later.

Both engines can be real. Neither is guaranteed.

Forced savings is powerful because it happens automatically if you keep making payments. But it's also illiquid. If you build $80,000 of equity and then lose your job, you can't swipe equity at the grocery store. Equity is not an emergency fund. It's a balance-sheet number tied to a building.

Appreciation is the part people love to talk about at barbecues. It's also the part that can disappoint in long stretches. Housing markets can stagnate. They can drop. They can recover slowly. They can vary wildly by neighborhood. Buying a home is not buying "the housing market." It's buying one specific property with one specific resale audience.

If buying makes you financially fragile—if it pushes your fixed costs to the edge—then the "investment" argument is upside-down. An investment that increases the probability of high-interest debt, missed retirement contributions, or a forced sale is not wealth-building in practice, even if it looks good on paper.

The tax break is real for some people, and imaginary for others

Another myth that props up the "renting is throwing money away" idea is the automatic assumption that homeownership comes with a big tax win.

Mortgage interest can be deductible in the U.S., but only if you itemize deductions, and the rules have limits. Many households take the standard deduction, meaning the mortgage interest deduction doesn't change their taxes at all. Even for itemizers, there are caps and conditions on what interest qualifies and how it's claimed.

The practical point isn't to memorize tax code. The practical point is: don't let a vague promise of "tax benefits" justify a payment you can't comfortably afford. If a home only works because you're counting on a tax break you haven't actually calculated, the deal isn't stable.

If you want the cleanest path, ask a tax preparer (or run tax software) with two scenarios: one where you rent, one where you buy. Use the same income. Compare the total tax. Then decide. Don't let a cultural slogan do your math.

Renting has its own risks---and you should name them, too

Renting isn't automatically "safer." It simply moves the risks around.

Rent increases. Depending on location, rent can rise fast. That can be destabilizing, especially if your income is flat. A fixed-rate mortgage can offer more payment stability over time (though taxes and insurance can still rise).

The landlord factor. A good landlord is invisible. A bad landlord can make your home feel temporary and stressful.

Lack of control. You can't remodel. You can't always have pets. You can't always stay as long as you want.

The key is to compare risks honestly. Homeownership risks are often larger but less visible day-to-day—until the day they become extremely visible. Renting risks are often smaller but more frequent—rent hikes, lease decisions, the feeling of impermanence.

The goal is to choose the set of risks you can handle without breaking your household.

A ``rent vs. buy" calculator you can do on a napkin

You don't need a complex model to avoid the most common mistake. Here's a simple way to get clarity:

1. Calculate your *all-in monthly cost to own*:

   P&I + property taxes + insurance + PMI + HOA + maintenance reserve

2. Subtract from that the part that builds equity *right now* (principal paid each month). Early in a mortgage, principal can be surprisingly small.
3. What's left is your *effective cost of owning*—the ownership version of "rent."

Then compare that effective cost of owning to rent.

If the effective cost of owning is much higher than rent, you are paying a premium for ownership—maybe for stability, maybe for a better living situation, maybe because you value having your own place. Paying a premium can still be a valid choice. The question is whether your household has the cash-flow strength to pay that premium without pushing you into a fixed-cost corner.

The overlooked move: ``Buy later" can be a powerful plan, not a failure

Some households treat renting as a waiting room you should exit as fast as possible. In reality, renting can be a strategic stage—especially if it's used to build the exact ingredients that make homeownership safe:

- a larger down payment (reducing PMI and monthly payment),
- a stronger emergency fund (so repairs don't become credit card debt),
- a lower debt load (so housing doesn't crowd out everything else),
- and a clearer time horizon (so you aren't buying right before a move).

"Buy later" isn't procrastination if it reduces your fragility ratio. It's a plan that protects your future self from having to solve a housing problem during a life problem.

A decision rule that prioritizes stability over status

If you want one rule that cuts through the noise, make it this:

> If buying pushes your fixed costs so high that one income loss would force debt, you're not buying a home-you're buying a fragile lifestyle.

That doesn't mean you must avoid buying. It means your purchase price, down payment, and monthly ownership stack need to be sized for the life you actually live, not the life you can maintain only when everything goes perfectly.

The most underrated luxury in housing isn't quartz countertops. It's the ability to say, "We can handle a rough year," and actually mean it.

Actionable takeaway: run the ``all-in" test before you fall in love with a listing

Before you tour another house (or renew another lease), run one honest comparison with your real numbers:

1. Get a realistic estimate of *PITIA* for the specific price range you're considering.
2. Add a maintenance reserve line item (start with 1% of home value per year divided by 12, then adjust for the home's age and condition).
3. Compare that total to your rent, *and* ask what happens if household income drops 20% for six months.
4. If the purchase only works when everything goes right, change the purchase price, increase the down payment, or keep renting while you strengthen your base.

The best housing decision is the one that lets you sleep—not because you own the walls, but because your life inside those walls isn't balanced on a knife edge.

3.3 Mortgages and Home Equity Risks

The most misleading word in personal finance might be *home.*

Not because homes are bad—they're often the center of a family's life—but because the word carries a warm glow that makes the risk feel softer than it is. A mortgage is not a vibe. It's a long contract that says, "No matter what happens—job loss, illness, recession, childcare chaos—we will produce this payment on schedule." When people get in trouble with housing, it's rarely because they didn't understand that a house costs money. It's because they didn't understand how *rigid* the costs are, and how *illiquid* the payoff can be.

Home equity looks fantastic on paper. It can also be useless at the exact moment you need help.

``House poor" isn't about being irresponsible. It's about being cornered.

Being house poor doesn't mean you bought a mansion with a fountain. It means your housing payment (and everything welded to it) eats so much of your cash flow that the rest of your life has to squeeze around it.

It shows up in small, humiliating ways before it shows up in big ones. You stop replacing worn-out shoes because "we'll do it next month." You keep a credit card balance because the statement closes before the next paycheck. You dread the dentist because you're not sure what insurance won't cover. You can't afford to take the risk that might improve your career—switching jobs, going back to school, moving for a better opportunity—because the house is a concrete anchor tied to your ankle.

House poor is a cash-flow condition, not a character flaw. Plenty of disciplined people become house poor the same way disciplined people can become sleep-deprived: by taking on a commitment that leaves no margin for being human.

A home is a leveraged asset. Leverage is powerful---and impatient.

A simple way to describe a typical home purchase is: you buy a large asset using a small amount of your own money and a large amount of borrowed money.

That borrowed money is leverage. Leverage can amplify gains (if the home rises in value, your equity can rise fast). It can also amplify stress. Your lender doesn't care what Zillow says your home is worth. The lender cares whether your payment arrives.

Leverage also changes the stakes of ordinary life events. If you rent and your income drops, you may be able to negotiate, downsize, or move when the lease ends. If you own with a tight budget, an income drop becomes a race between your savings account and the due date.

The problem isn't the existence of a mortgage. The problem is a mortgage sized so aggressively that your life can't breathe around it.

Home equity: impressive on a balance sheet, stubborn in real life

Home equity is the difference between what your home could sell for and what you owe on it. It's real wealth. It's just not *liquid*.

Liquidity means "spendable quickly without taking a big loss." Cash is liquid. A home is not. Turning equity into spendable money typically requires one of these moves:

- *Sell the home* (slow, expensive, disruptive, and sometimes impossible on your timeline).
- *Borrow against the home* (cash-out refinance, home equity loan, or HELOC), which creates a new payment and new risk.
- *Downsize* (which still requires selling, buying, moving, and paying transaction costs).

This is why home equity can be both true and unhelpful. You can be "worth" $300,000 on paper and still be one missed paycheck away from panic because the wealth is trapped in drywall.

There's also a psychological trap: equity feels like a safety net, so people allow their cash reserves to run thin. Then a repair hits—roof, HVAC, plumbing—and suddenly the "wealthy" household is juggling credit cards.

A concrete case: the equity mirage

Consider Tasha and Devon, who bought a home five years ago. Their neighborhood did well. Similar homes are selling for much more. They feel relieved; they tell themselves, "At least we have equity."

Then Devon's employer changes commission structure and his income becomes choppier. One month is great. The next is ugly. The mortgage is steady either way.

They look at their situation:

- They have $160,000 in home equity (on paper).
- They have $4,000 in their emergency fund (in reality).

- They need $8,500 to replace an aging HVAC system (also in reality).

The equity doesn't fix the HVAC. Cash fixes the HVAC.

Could they borrow against the equity? Possibly. But borrowing takes time, paperwork, credit approval, appraisal, fees, and then it produces a new payment. And if income is already unstable, adding another required payment is like patching a leak by raising the water level.

So they do what many households do: they put the HVAC on a credit card "temporarily," promising themselves they'll pay it off when things normalize. That's how home equity turns into a mirage—comforting from a distance, useless up close.

The moral isn't "never buy" or "never use credit." The moral is that a home with thin cash reserves is not a stable asset. It's a beautiful, expensive source of surprise invoices.

The payment you can afford is not the payment you should choose

There's a reason lenders talk about what you can "qualify" for. The word is revealing. Qualifying is not thriving.

A mortgage approval is based on formulas and averages. Your household is not an average. You have real childcare costs, real commuting costs, real medical needs, real family obligations, and real stress limits. A lender doesn't price in your desire to sleep at night.

This is where "house poor" is born: a household takes the maximum payment that the system will allow instead of choosing a payment that leaves margin for normal life. The extra bedroom becomes a monthly tax on your future flexibility.

A useful way to think about it is: a mortgage payment is not just a number. It's a claim on your time. The higher it is, the fewer choices you have when your job is miserable, when a parent needs help, when a child needs more support, when you simply want to change direction.

Fixed-rate vs. adjustable-rate: the difference between a steady bill and a moving target

A fixed-rate mortgage is boring in the best way. The interest rate stays the same for the life of the loan, which helps keep the principal-and-interest payment stable. (Taxes and insurance can still rise, which is its own headache, but at least the rate isn't another variable.)

An adjustable-rate mortgage (ARM) can start with a lower rate, which is why it can look like a clever hack. The catch is in the name: it adjusts.

When an ARM adjusts, the payment can increase—sometimes sharply. The details matter: how long the introductory period lasts, how often it adjusts after that, and how much it's allowed to change. Those limits are called *rate caps*. A rate cap doesn't eliminate risk; it simply defines the maximum size of the surprise.

Here's the practical danger: many households buy with an ARM because the initial payment is the only way the home "works." That means the budget is already tight *before* the payment risk kicks in. If your plan depends on refinancing later, you're making a second bet: that interest rates will cooperate and your income and credit will be strong at the exact moment you need them to be. Plans that require perfect timing are not plans—they're hopes with paperwork.

ARMs can be reasonable tools for some people, especially with strong cash reserves and a clear, short time horizon. They are a sharp tool, though, and sharp tools demand respect.

The ``payment shock" question you should insist on answering

When a mortgage payment is going to be the biggest fixed cost in your life, you deserve clarity that's almost rude.

If you're considering an ARM, ask for this in plain language:

> *"Show me the highest possible monthly payment I could ever face under this loan's terms."*

Not "an estimate." Not "probably." The maximum, given the caps.

Then ask the only question that matters: if that maximum payment happened during a job disruption, could your household handle it without running up expensive debt?

If the answer is no, then the ARM isn't a bargain. It's a cliff with a nice view.

The quiet risk inside ``low down payment" loans: being fragile early

Low down payment options can be the difference between buying and not buying, and for some households they're a valid bridge. But they come with a season of extra fragility that many buyers underestimate.

Early in a mortgage, three things often happen at once:

- Your monthly payment is high relative to your income.
- Your equity is thin.
- Your principal paydown is slow (especially at the start), meaning equity builds gradually.

If something goes wrong early—job loss, medical event, a needed move—you have fewer options. Selling is expensive. Equity may not cover selling costs. Refinancing may be impossible. That's not doom; it's just the physics of a leveraged purchase with high transaction costs.

This is why "Can we make the payment?" is the wrong affordability test. A better test is: "How many things can go wrong before this house forces our hand?"

HELOCs: the tempting way to turn equity into cash---and the risk of turning your home into collateral for everything

A home equity line of credit (HELOC) is a revolving line of credit secured by your home. In plain terms, it's like a credit card with a lower interest rate and a much nastier consequence if you fall behind.

People use HELOCs for renovations, debt consolidation, college costs, and "just in case." Some uses are reasonable. Some are financial self-sabotage in a friendly font.

The risk isn't that a HELOC exists. The risk is what it does to your household's fixed-cost structure and to your downside risk. Because the HELOC is secured by the home, failing to keep up can put the roof itself at risk. That's qualitatively different from a typical unsecured credit card.

There's also the "normalization" problem: once a HELOC is open, it can feel like part of your emergency fund. It isn't. An emergency fund is money you already have. A HELOC is money you can borrow—with a payment that can rise, terms that can change, and approvals that can tighten when the economy gets ugly.

A surprisingly common pattern looks like this:

- Household buys a home with a tight budget.
- Repairs and life costs show up.
- They use a HELOC to smooth the bumps.
- The HELOC payment becomes another fixed cost.
- Now the household is more house poor than before, just with nicer cabinets.

If you use a HELOC, treat it like a power tool: useful, dangerous, and never something you leave on the floor around kids.

``But the interest is deductible" is not a plan

Tax myths have a way of sticking to housing like gum on a shoe.

Mortgage interest deductibility depends on itemizing deductions, and there are caps and rules. Interest on home equity borrowing can also have restrictions—particularly if the borrowed funds weren't used to buy, build, or substantially improve the home. The result is that many households either get no benefit or a smaller benefit than they assumed, especially when the standard deduction is the better choice.

Even when there is a tax benefit, it works like a discount, not a rebate. Spending a dollar to save a fraction of a dollar on taxes is still spending a dollar.

If a loan only feels affordable because you're counting on a tax angle you haven't verified, the affordability is imaginary.

Choosing mortgage terms that protect cash flow (instead of feeding your pride)

There's a certain kind of financial pride that shows up around mortgages: shorter terms, aggressive extra payments, the dream of being "debt-free" fast. That can be a strong goal. It can also become a trap if it starves your household of liquidity.

The key tradeoff is simple:

- Paying extra toward the mortgage increases equity.
- Keeping cash available increases resilience.

Equity helps you in a stable future. Cash helps you in an unstable present.

A household with a huge emergency fund and stable jobs can afford to be aggressive. A household with variable income, high childcare costs, medical complexity, or a fragile fixed-cost structure often benefits more from flexibility than from speed.

One practical strategy is to separate the *ability* to pay extra from the *obligation* to pay extra. You can always choose to make additional principal payments in good months. You cannot easily choose to lower your required payment in bad months.

That's why "opt-in aggressiveness" tends to beat "locked-in aggressiveness" for many two-income households.

The homeowner's resilience stack: cash first, then equity

If you want housing to be stabilizing instead of stressful, build the supports in the right order.

1. *Keep the monthly payment survivable on one income for a while* (even if it's uncomfortable to admit you might need that).
2. *Hold real liquidity* (cash reserves for repairs and income gaps).
3. *Then focus on growing equity faster* (extra principal, improvements that make sense, longer-term planning).

This order frustrates people because it feels "slow." But it's how you avoid the classic disaster: living in an expensive asset while financing daily life on high-interest debt.

The goal isn't to worship cash. The goal is to ensure that your home doesn't turn every ordinary surprise into a borrowing event.

Actionable takeaway: do a ``liquidity stress test" before you accelerate payoff or borrow against equity

Pick a quiet hour and run this stress test using your real numbers:

1. Write down your *required* monthly housing cost: mortgage principal and interest *plus* taxes and insurance (escrow) *plus* HOA *plus* PMI, if any.
2. Add the other fixed costs that can't be dodged quickly (minimum debt payments, childcare contracts, required transportation).
3. Now simulate a realistic shock: one income drops by 30–50% or disappears for 3 months.
4. Answer two questions honestly:
    - How many months can we pay every required bill without borrowing?
    - If we needed $8,000 tomorrow for a repair, could we pay it without turning the house into collateral or carrying a credit card balance?

If the answers are uncomfortable, don't solve it with denial or another loan. Solve it with structure: reduce fixed costs where possible, rebuild liquidity, and treat home equity as long-term wealth—not as a checking account with drywall around it.

3.4 Budgeting Methods that Work

Budgets have a branding problem.

The word itself sounds like punishment—a financial version of being grounded. Many households try budgeting exactly once, do it the strict way (track every purchase, shame yourself for every slip), and then decide budgets don't work for them. What usually doesn't work is a budgeting *method* that fights your personality, your schedule, and the reality that life is messy.

A budget isn't a moral document. It's a tool for turning cash flow into choices. And once fixed costs are trimmed as far as they reasonably can be, a workable budget is what keeps those gains from evaporating the moment life gets busy again.

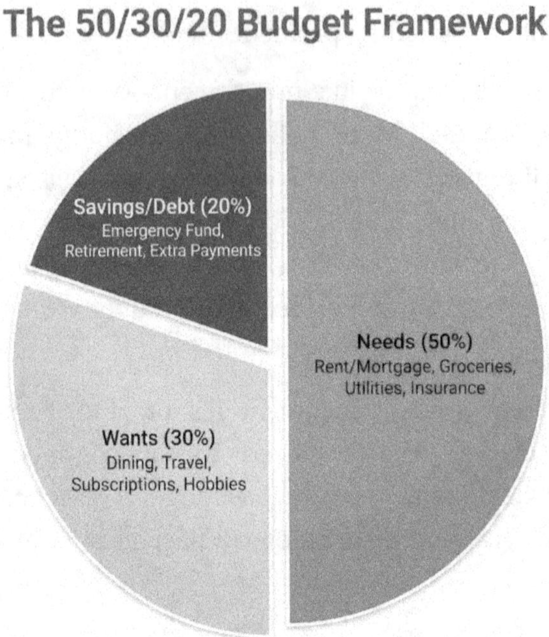

The real purpose of a budget is not to prove you have willpower. It's to make sure the money you *care about*—your stability, your savings, your debt payoff, your family goals—doesn't get eaten alive by the money you spend in a hurry. Most spending "mistakes" are just decisions made at the wrong time of day: tired, hungry, busy, stressed, or trying to buy five minutes of peace.

So the question isn't, "Can you stick to a budget?" The question is, "Can you find a system that makes your best decisions the default?"

A budget that works has two jobs: guardrails and truth

A workable system does two things at once.

First, it provides guardrails. Not concrete walls, not a straightjacket—guardrails. You can still steer. You just can't accidentally drive off a cliff because nobody told you the road turned.

Second, it tells the truth. Many households operate with an optimistic fog: "We're doing fine, we're just busy." A budget replaces fog with numbers. Not to judge you, but to stop you from being surprised by your own life.

That truth-telling function is especially valuable for irregular expenses—the ones that show up like rude houseguests. Car repairs. Back-to-school costs. Holiday travel. Annual subscriptions. Medical bills that land in the same month as a birthday party and an out-of-town wedding. If your budget only handles "normal months," it's not a budget. It's a wish.

Method #1: The 50/30/20 rule (the friendly framework that prevents silent drift)

The 50/30/20 rule is popular for a reason: it's simple enough to remember on a bad day.

It suggests dividing your after-tax income into three broad buckets:

- *Needs (50%):* the basics you can't easily skip—housing, utilities, groceries, insurance, minimum debt payments, transportation to work.

- *Wants (30%):* the life-enhancing stuff—restaurants, entertainment, trips, hobbies, upgrades.
- *Savings/Debt (20%):* emergency fund contributions, retirement investing, and extra payments beyond minimums.

This framework is not a law of physics. It's a diagnostic. If your needs are at 65%, you're not "bad." You're constrained. If your wants are at 40%, you're not "wasteful." You may be stressed and buying relief. The categories help you see where reality is pulling you.

The quiet superpower of 50/30/20 is that it stops lifestyle inflation from sneaking into the walls. When income rises, the default temptation is to upgrade needs (bigger place, newer cars) and upgrade wants (more convenience spending). The 50/30/20 framework gently insists: *your future deserves a slice too.*

How to use 50/30/20 without turning it into a guilt chart

People misuse this rule by treating the percentages as grades. That's not the point. Use it like a health check:

- If *Needs* are too high, the problem is often fixed costs (housing and transportation are the usual suspects). The best "budgeting" move might be renegotiating insurance, changing a car situation, or choosing a different housing plan when the lease ends.
- If *Wants* are too high, you can often reclaim a surprising amount of money by identifying the handful of cate-

gories that spike when you're tired (delivery food, convenience shopping, weekend "treat" spending that became a habit).

- If *Savings/Debt* is too low, don't start by promising yourself you'll "try harder." Start by automating a small amount and gradually increasing it. Humans are better at adapting than at depriving.

Also, if your household income is volatile, percentages can be hard. In that case, use 50/30/20 as an average over several months, not as a strict rule for every pay period.

Method #2: Zero-based budgeting (the method for people who hate surprises)

Zero-based budgeting sounds intense, like you need a visor and twelve spreadsheets. The concept is simpler: every dollar of income is assigned a job before the month begins, so that "extra money" doesn't quietly disappear.

The "zero" doesn't mean you spend everything. It means you plan everything. Savings is a category. Debt payoff is a category. A future car repair is a category. When the plan is done, income minus planned categories equals zero because every dollar is accounted for.

This method is excellent for households who feel like money leaks out of their life in tiny invisible holes. It's also excellent for households with irregular expenses, because it forces you to acknowledge them ahead of time.

Zero-based budgeting is less about control and more about reducing the number of decisions you have to make while tired. Decision fatigue is expensive.

The missing ingredient: "true expenses" and sinking funds

The reason many budgets fail is not that people can't follow rules. It's that the budget pretends irregular costs don't exist, then acts offended when they arrive.

A more realistic approach is to treat predictable irregular expenses as monthly line items. This is often called a sinking fund: you set aside a little each month so the eventual bill doesn't become an emergency.

Examples:

- Car maintenance: $100/month
- Gifts/holidays: $75/month
- Medical out-of-pocket: $150/month
- Travel to see family: $80/month
- Home repair fund (if you own): a real number, not a vibe

The goal isn't perfection. The goal is fewer "How did that happen?" moments.

A concrete case: the household that stopped "budgeting" and started assigning jobs

Consider Chris and Nia, a two-income household with a toddler. They have decent salaries, but they keep landing on their credit card at the end of the month. They don't feel extravagant; they feel busy. They try cutting restaurants and subscriptions, and it helps for two weeks. Then the daycare fundraiser hits, the car needs tires, someone gets sick and they order food all week, and they're back where they started.

What changed things wasn't a stricter attitude. It was a clearer plan.

They switched to a zero-based approach and built three sinking funds:

- "Kid stuff we forget about" (clothes, school events, birthday parties): $120/month
- "Car reality" (maintenance and repairs): $150/month
- "Health surprises" (copays, prescriptions, random bills): $100/month

They also made one uncomfortable choice: they stopped using their credit card as a shock absorber. Instead, they kept a small buffer in checking so a weird month didn't instantly become debt.

In the first month, nothing looked magical. In the third month, something shifted: the same annoyances happened, but they didn't feel like emergencies. Their finances got calmer, not because life got cheaper, but because the budget stopped pretending life was predictable.

Zero-based budgeting didn't make them "better." It made them less surprised.

Method #3: The anti-budget (for households who need simplicity, not micromanagement)

Some people would rather floss with barbed wire than track every category. If that's you, the anti-budget can be a lifesaver.

The anti-budget flips the normal idea. Instead of tracking spending, you automate the important things first and then spend the rest without guilt.

The basic structure:

- Money automatically goes to savings (emergency fund, retirement, sinking funds) on payday.
- Minimum debt payments are automated.
- Fixed bills are automated.
- The remaining money is your "spend" money for the month (or week).

The anti-budget is not "no rules." It's "fewer rules in the right places." It works best when you have enough margin that you don't need to squeeze every category, and when your biggest risk is not overspending in one specific place but generally drifting.

If you tend to rebel against detailed budgets, the anti-budget respects that. It says: we'll lock in the priorities, then let you live your life.

How to choose the right method: match the system to the failure mode

Budgeting method isn't about virtue. It's about friction.

Different households have different failure modes:

- If you don't know where the money goes and you feel blindsided each month, zero-based budgeting is your friend.
- If you need a simple sense of balance and a way to catch lifestyle inflation, 50/30/20 is a strong starting point.
- If you can't stand detailed tracking and you mostly need to ensure saving actually happens, the anti-budget is often the most sustainable.

Notice what all three methods share: they create a protected space for your future. The differences are in how much detail you need to make that protection real.

The secret sauce: automation beats motivation

Motivation is a spark. Automation is a furnace.

One of the most reliable ways to increase saving over time is to commit future raises to saving before you have a chance to get used to the bigger paycheck. It's psychologically elegant: you don't feel like you're losing money, because you never built your lifestyle around it.

This is the logic behind "save more tomorrow" approaches: set a plan now that automatically increases contributions later. Many employers allow automatic increases to retirement contributions each year or each raise cycle. You can also do it manually with a calendar reminder tied to performance review season.

If you're waiting to feel ready to save more, you'll be waiting through a lot of normal life: busy seasons at work, kids' needs, travel, health stuff, burnout, holidays. A system that saves for you while you're distracted is not laziness. It's realism.

A budget meeting that doesn't ruin your relationship

In a two-income household, budgeting fails for a sneaky reason: it becomes a proxy war about values, exhaustion, and fairness.

One person feels like the responsible adult. The other feels policed. Or one person does all the financial labor and starts to resent it. Or both people avoid the conversation because it always turns into a fight about takeout.

A better approach is to keep the "money meeting" short and concrete:

- Look at cash flow, not character.
- Agree on one number that matters this month (a savings target, a debt payoff target, a fixed-cost reduction, a buffer level).
- Decide what you're saying no to in order to say yes to that number.

Then stop. You're building a habit, not staging a courtroom drama.

Small tactics that keep the system from collapsing in real life

Even the best framework can fall apart in the face of a chaotic month. A few practical moves help:

Weekly spending checkpoints. If you only check once a month, you discover problems when it's too late. A five-minute weekly look keeps you from accidentally spending the rent money on a week of "we deserve this."

A separate account for sinking funds. If irregular-expense money sits in the same checking account as your daily spending, it will get spent. Not because you're irresponsible, but because money is a liquid you can't label with a Sharpie. Separate it.

A "minimum viable budget" for rough seasons. Some months you can track every category. Some months you're just trying to survive. In rough seasons, fall back to three protected items: keep the bills current, keep the savings automation running if possible (even at a reduced level), and stop new debt. That's enough to prevent long-term damage.

Actionable takeaway: build your budget in one hour, then let it run

Set a timer for 60 minutes this week and do the following:

1. Choose your method: 50/30/20, zero-based, or anti-budget. Pick the one you will actually use, not the one that sounds impressive.
2. Identify your non-negotiables: fixed bills and minimum debt payments.
3. Create at least one sinking fund for a predictable irregular cost that has been hurting you (car repairs, medical, gifts, school).
4. Automate one priority transfer on payday (even $25): emergency fund, retirement, or extra debt payment.

5. Pick one weekly check-in time (10 minutes) to look at balances and upcoming bills.

A working budget doesn't make you feel constrained. It makes you feel *harder to surprise*. And in a household where the biggest risks come from a few large, rigid obligations, being harder to surprise is a kind of wealth.

3.5 Saving Rates and Targets

The most important number in your financial life is not your salary. It's not your credit score. It's not even your net worth.

It's the *gap*—the distance between what comes in and what goes out.

That gap is your household's engine. It pays for emergencies without panic. It buys down debt before interest eats you. It funds retirement, college, a sabbatical, a career change, a move closer to family, a new business, or simply the ability to breathe. If your life were a house, the gap is the foundation you don't post on social media. If it's strong, the whole structure can take a hit. If it's cracked, every new expense feels like a threat.

What makes the gap such a powerful metric is that it works no matter what you earn. Two households can make $80,000 and live in different universes depending on how wide the gap is. One has options. The other has obligations.

Saving rate: the thermostat, not the scoreboard

A saving rate is simply the percentage of your income that you keep for future you. Depending on your goals, *saving* can include retirement contributions, emergency fund deposits, sinking funds for upcoming expenses, and extra payments on high-interest debt (because paying off expensive debt is a guaranteed return).

People often treat saving rate like a badge: higher is better, therefore anything lower is failure. That mindset is a fast track to quitting. Saving rate is better used like a thermostat. You set a target based on the kind of life you want and the kind of risk you're carrying, then you adjust it over time.

In a two-income household, the thermostat metaphor matters. Some years are expensive: childcare peaks, a parent needs help, medical costs flare up, an income becomes unstable. A thermostat can adjust without shame. A scoreboard just tells you you're losing.

The dirty secret: *saving more* is usually a fixed-cost story

Many households think their saving problem is a self-control problem. It's often a structure problem.

If you have a big monthly nut—high housing cost, high car payments, high debt minimums—then saving becomes something you try to do *after* life is paid for. That's the worst possible position for saving. "After" is where money goes to disappear. It gets eaten by surprise expenses and convenience spending, because your brain is exhausted from managing the basics.

Widening the gap is why the earlier work of reducing fixed costs matters. When your fixed obligations drop, your saving rate can rise without turning your life into a bleak endurance contest.

The goal isn't to become a monk. The goal is to stop being surprised by your own bills.

Pay Yourself First: turning saving into a bill you can't ``forget"

The simplest behavioral trick in personal finance is also one of the most effective: pay yourself first.

That means savings happens automatically, at the beginning of the month or on payday, before you have a chance to spend the money accidentally. Not because you're weak, but because you're human. If the money is in your checking account, it will be assigned jobs by whatever is loudest that week: a birthday party, a broken appliance, a stressful Tuesday.

Paying yourself first transforms saving from a mood into a system. It's the difference between saying, "We should save more," and saying, "Saving is part of our cost of living."

You can do this with:

- automatic transfers to a high-yield savings account,
- payroll contributions to retirement accounts,
- automated extra payments to debt (especially high-interest debt),
- or automatic deposits into sinking funds for predictable irregular expenses.

If it isn't automatic, it isn't real in the months when life is chaotic. And those months are not rare. They're the default.

How much should you save? Start with two targets: resilience and retirement

A saving target that tries to do everything at once usually does nothing. It's more effective to set two distinct targets with different purposes:

1. *Resilience savings:* cash and near-cash that prevents a rough month from turning into debt.
2. *Long-term savings:* retirement investing and other long-horizon goals.

They work together, but they solve different problems.

Resilience savings is about staying on your feet. Long-term savings is about building wealth. If you only do long-term savings without resilience, you can end up raiding retirement accounts or racking up credit card debt when life hits. If you only do resilience without long-term savings, you feel safe today but leave future you with a painful bill.

The right balance depends on how fragile your household is, how stable your income is, and how close you are to retirement. But the structure is the same: one bucket for "don't fall," one bucket for "build."

A realistic retirement saving benchmark (and why it's not a commandment)

A common mainstream guideline is to save on the order of roughly 12%–15% of pay per year for retirement when you include employer contributions. For many households, that's a solid long-term target because it tends to support a reasonable retirement without requiring extreme lifestyle changes.

But guidelines can be dangerous when they're heard as judgment. If you're currently saving 3%, being told "15%" can feel like being told to dunk a basketball when you can't touch the rim. You don't need a lecture; you need a ramp.

The most useful way to treat a benchmark is to ask: *What would it take to move one notch closer, and what would we be giving up?* Then repeat over time.

The ladder approach: saving rate increases that don't ruin your life

A saving-rate ladder is a plan to raise your saving rate gradually, in steps, rather than by trying to jump from "not saving" to "super saver" overnight.

Here's a simple version that works for many households:

- Pick a starting rate you can live with (even 1%–2%).
- Automate it so it happens without negotiation.
- Every time you get a raise, divert half of the raise to saving until you reach your target.

This is powerful because it harnesses adaptation. Humans adjust quickly to a slightly higher saving rate if it happens at the same time as an income increase. You don't feel deprived because your take-home still rises. You just don't inflate your fixed costs to match.

This approach is closely aligned with "save more tomorrow" style strategies: commit now to saving more later, tied to raises. It's one of the cleanest ways to widen the gap without relying on constant willpower.

A concrete case: the couple who stopped trying to ``cut spending" and started protecting the gap

Jules and Priya are both working, and their income is decent. They're not reckless, but money never seems to stay put. They decide to "save more," and for a month they do: fewer dinners out, fewer impulse purchases. Then life happens—a friend's wedding, a medical bill, and a stretch of exhausting work that turns into delivery food. The savings goal evaporates. They conclude they're not disciplined.

What actually changed their trajectory was one structural move and one behavioral move.

Structural: they refinanced their auto insurance and switched phone plans, saving $160/month. Not glamorous, but permanent.

Behavioral: they set an automatic transfer of $200 to savings the day after payday and an automatic 1% increase in retirement contributions every six months until they hit 12%. They also created a "wedding and travel" sinking fund of $100/month because, in their lives, that wasn't a surprise—it was a pattern.

In the first month, they still ordered delivery sometimes. They still had a chaotic week. The difference was that the saving happened anyway. The system widened the gap quietly, while their habits caught up more slowly.

That's the point: build a plan that works even when you're not at your best.

The hidden enemy of saving: not spending, but *spending you can't easily undo*

Saving is easiest when your expenses are flexible. It's hardest when your expenses are locked in.

A household can have a great income and still struggle to save because fixed costs consume the majority of the paycheck. In that situation, you don't have a "saving problem." You have a "commitment problem." You can't out-budget a mortgage that is too large for your life, or two car payments that eat your margin, or debt minimums that are the price of past emergencies.

This is why the gap is so sensitive to fixed costs. When fixed costs rise, saving becomes optional. When fixed costs fall, saving becomes possible.

What counts as saving when you're digging out?

One reason households get discouraged is that they're making progress but it doesn't look like "saving."

If you're carrying high-interest debt, extra payments are a form of saving because they reduce future required outflows. They widen the gap later. Paying off a credit card balance at 24% APR is not morally virtuous; it's financially lethal to keep it around. Eliminating it is a guaranteed, risk-free "return" equal to the interest rate you avoid.

Similarly, building a small emergency fund while paying down debt can prevent a single surprise from throwing you back into the same hole. The balance depends on your situation, but the mindset is important: a household in rebuild mode is still saving, even if the first "savings" shows up as fewer minimum payments.

Targets that don't collapse under real life

Good targets are specific enough to guide action and flexible enough to survive a messy month.

A practical way to set saving targets is to define three layers:

- *Floor:* the minimum you will save no matter what (even $25/week). The floor keeps the habit alive.
- *Base:* the normal target in an average month.
- *Stretch:* what you do in good months when you have extra margin (bonuses, extra paychecks, unusually low expenses).

This approach prevents the all-or-nothing cycle where one rough month makes you quit entirely. The floor is especially important for two-income households with variable schedules or incomes. A plan that only works in perfect months is not a plan.

A quick calculation: the ``gap number" that should make you nervous (or proud)

If you want a single number to track that's more emotionally honest than net worth, try this:

Gap Number = Monthly Take-Home Pay – Monthly Fixed Costs

This is not your full spending. It's your margin before life happens.

If the gap number is small, your household is fragile even if your income is high. If the gap number is large, you can handle surprises without debt and you can increase savings aggressively when you choose.

Track the gap number for three months. If it's shrinking, something is quietly inflating—often fixed costs or "subscription creep." If it's growing, you're building real power.

Actionable takeaway: set a saving-rate ladder and automate the first rung today

Do one concrete thing in the next 30 minutes:

1. Choose a starting saving rate you can tolerate right now (1%, 3%, 5%—whatever won't cause a backlash in your life).
2. Automate it so it happens on payday: retirement contribution increase, automatic transfer to savings, or an automatic extra debt payment.
3. Write a one-sentence ladder rule: *"Half of every raise goes to saving until we reach X%."*
4. Put a calendar reminder on the date of your next raise cycle (or pick two dates per year) to increase the rate by 1%.

The gap doesn't widen because you had an unusually disciplined month. It widens because you built a system that makes saving the default and lifestyle inflation the exception. When that system is in place, wealth-building stops being a heroic act and starts being the quiet thing your household does in the background while you live your actual life.

3.6 Building the Emergency Fund

Nobody wakes up excited to build an emergency fund. There's no "emergency fund aesthetic." You can't drive it, wear it, or show it to your neighbors. It just sits there, quietly refusing to be interesting.

And then one day it becomes the most beautiful money you own.

The emergency fund is not an investment strategy. It's a shock absorber. It's the difference between a crisis that stays a crisis and a crisis that turns into a long, expensive, relationship-straining season of debt. In a two-income household, it's also a kind of emotional permission slip: the ability to say, "We can handle this," and mean it without doing mental gymnastics.

The tricky part is sizing it. Too small, and you're one bad month away from borrowing. Too large, and you may be starving other goals. The right size isn't a moral number. It's a risk number.

Emergency fund money has one job: keep problems from multiplying

A broken transmission is a problem. A broken transmission plus a maxed-out credit card plus a missed mortgage payment plus a fight about whose fault it is—that's a cascade.

Emergency funds exist to prevent cascades. They turn sharp, expensive surprises into blunt, manageable ones. Instead of panicking, you pay. Instead of borrowing at ugly interest rates, you use cash. Instead of choosing between groceries and the electric bill, you keep your life intact while you figure out what's next.

This is why emergency funds feel "unproductive" right up until the moment they prevent a financial chain reaction.

Two incomes don't always make you twice as safe

Two incomes can be safer, especially when they're independent and stable. But many households accidentally structure their lives so that both incomes are required to keep the bills current. That's the danger zone: the household looks strong because the total income is high, but it's actually more exposed because the monthly nut has expanded to match.

Two incomes can also be correlated in ways people don't expect. If you work in the same industry, a downturn can hit both jobs. If one income supports the other (childcare logistics, commuting, a shared vehicle), a disruption in one person's schedule can disrupt both. And if your household relies on overtime, bonuses, or commission, your "income" may be partly weather-dependent.

The emergency fund is what lets you survive the period where the world is figuring out what it wants from your paycheck.

The most useful definition: your emergency fund is time

People often ask, "How many months should I save?" That question is right, but it's missing the key concept. What you're really buying is *time*.

Time to job search without desperation. Time to negotiate instead of accepting the first offer out of fear. Time to deal with a family medical issue without detonating your finances. Time to repair your car without turning it into a high-interest loan.

In other words: time to make good decisions.

The emergency fund is not there to maximize returns. It's there to keep you from making low-quality decisions under pressure.

A baseline target (and why it's only a baseline)

A common guideline is to keep about 3 to 6 months of living expenses in a liquid, easily accessible account. That's a reasonable starting point for many households. It's not a universal law, because households don't have equal risk.

The range exists for a reason:

- Some households have stable, salaried jobs in resilient fields and strong support networks.
- Others have variable income, seasonal work, commissions, or industries prone to layoffs.

- Some households have dependents, health complexity, or caregiving responsibilities that make disruptions more likely and more expensive.

The right number is the one that matches your household's actual exposure.

A concrete case: the ``we're fine" family that wasn't

Leah and Omar both work. Their combined income looks solid on paper. They own a home, have two kids, and generally feel like responsible adults.

Then Omar's company announces layoffs. He gets a severance package that sounds decent—until they realize how quickly fixed costs eat it. At the same time, Leah's job becomes more demanding, because half her team is out sick and the workload piles up. The kids' after-school program is closed for two weeks due to staffing issues. Their schedule collapses. Omar's job search becomes the full-time job, but the logistics of childcare, commuting, and interviews become a constant puzzle.

They have $6,000 in savings. It feels like a lot until they do the math:

- Mortgage, taxes, insurance: $3,200
- Childcare and after-school care: $1,400
- Car payment and insurance: $850
- Minimum debt payments: $450
- Utilities and phones: $450

Their essential nut is around $6,350 before groceries. Their savings is less than one month of essentials.

So what happens? They do what many households do: they float. A credit card becomes the bridge. Then the bridge becomes a burden. Interest turns a short-term problem into a longer-term drag. The stress spills into everything: sleep, work performance, relationships.

The painful part is that Leah and Omar weren't reckless. They were under-buffered. The emergency fund wasn't about "being prepared for the apocalypse." It was about being prepared for an extremely normal event: a job disruption plus life complications.

Sizing your emergency fund with a simple formula

Instead of picking a number you've heard other people say, size your fund based on two inputs:

1. *How hard would it be to replace your income?*
2. *How many people depend on that income staying steady?*

Here's a practical way to turn that into a number without creating a spreadsheet empire.

Step 1: Define "essential monthly expenses"

This is the monthly amount you must cover to keep your life stable. It is not your current spending. It is the survival version:

- housing (rent/mortgage plus required insurance/taxes/HOA),

- basic utilities,
- groceries (not restaurants),
- essential transportation,
- insurance premiums,
- minimum debt payments,
- required childcare to work,
- basic medical needs.

If you've ever said, "We could cut back if we had to," this is where you define what that actually means. Don't make it heroic. Make it realistic. People fail at emergency plans when the plan assumes they'll suddenly become a different person under stress.

Call this number E (for essentials).

Step 2: Choose your runway length based on risk

Now choose a number of months, M, based on your household risk profile. A simple risk-based guideline looks like this:

- *3 months:* high job stability, no dependents, strong backup options.
- *6 months:* moderate stability and/or dependents, or one income is variable.
- *9 months:* lower stability, commission/seasonal work, single breadwinner, or multiple dependents.

- *12 months:* high volatility, specialized job market, health complexity, or a household that would be severely disrupted by an income loss.

This isn't about pessimism. It's about acknowledging that replacing income is not always quick, and when you have dependents, "quick" can still be too slow.

Step 3: Compute the target

$$\text{Emergency Fund Target} = E \times M$$

If your essential monthly expenses are $5,000 and you choose 6 months, your target is $30,000. If that number makes you feel dizzy, that's information. It tells you your essential expenses are high, your risk is high, or both.

Sometimes the best "emergency fund" move isn't saving faster. It's lowering E by reducing fixed costs. If your essentials are $7,000 per month, every month of runway is expensive. Lowering the nut makes every other safety strategy cheaper.

The emergency fund doesn't need to be built all at once

Large targets can trigger an all-or-nothing response: "We'll never get there, so why start?" That's a motivation trap.

Think in milestones:

- *The $1,000 buffer:* prevents small shocks from hitting credit cards.

- *One month of essentials:* a massive psychological shift; you can handle a job delay or a medical bill without immediate borrowing.
- *Three months:* enough to survive many disruptions.
- *Six months+:* true negotiating power.

Each milestone changes your behavior. When you have no buffer, you react. When you have a buffer, you plan.

Where to keep it: boring, liquid, and separate

An emergency fund has to be accessible quickly and safe from market swings. That means it's usually not invested in stocks. The goal is stability, not growth.

A high-yield savings account is a common choice. The key features you want:

- easy access within a day or two,
- FDIC or NCUA insurance (in the U.S.),
- separated from your daily checking so you don't spend it by accident,
- and ideally not tied to your debit card.

Separate doesn't mean inaccessible. It means you have to make a conscious decision to transfer and use it. That tiny bit of friction protects the fund from "emergencies" like a flash sale or a stressful week of takeout.

What counts as an emergency (so you don't quietly drain it)

The emergency fund is not for everything that feels urgent. It's for:

- job loss or income interruption,
- medical expenses that can't wait,
- essential home or car repairs,
- urgent travel for family emergencies,
- safety issues.

It is not for predictable irregular costs you could have planned for (holidays, annual subscriptions, routine maintenance). Those belong in sinking funds, not in the emergency fund, because raiding the emergency fund for predictable costs turns it into a revolving door.

A useful rule is: if it's likely to happen eventually, it's not an emergency. It's a category.

The relationship benefit: emergency funds reduce money fights

Many financial arguments are really arguments about fear.

When one partner feels like the household is one surprise away from disaster, every purchase feels threatening. When there's a real buffer, the emotional temperature drops. You can discuss tradeoffs without the background soundtrack of panic.

In two-income households, an emergency fund also changes the power dynamics in a healthy way. It can give someone permission to leave a toxic job, to take parental leave, to reduce hours temporarily, or to go back to school. That's not indulgence. That's using money to protect your family's long-term stability.

Actionable takeaway: do the ``one month" sprint and lock it in

If you want a concrete plan you can start today, do this:

1. Calculate your essential monthly expenses E (the realistic survival version).
2. Set your first target at *one month of essentials*, even if your ultimate goal is 6 or 9 months.
3. Automate a weekly transfer into a separate savings account. Pick a number that feels almost too easy (for example, $50/week). The goal is consistency.
4. Any time you get unexpected money (tax refund, bonus, gift, side income), send a fixed percentage (say 50%) to the emergency fund until you reach one month.
5. Once you hit one month, keep going—but pause for a moment and notice what just changed: you bought yourself time. Time is what turns a household from fragile to resilient.

An emergency fund is not proof that nothing bad will happen. It's proof that when something bad happens, you won't have to make it worse just to survive it.

Chapter 4

Eliminating Debt and Insuring the Plan

Imagine trying to fill a bathtub while the drain is wide open and the roof above you is leaking. That is the reality of building wealth while carrying high-interest debt and ignoring major risks. In this chapter, we stop the bleeding. We will plug the drain by attacking debt with psychological warfare, and we will patch the roof by setting up an 'iron dome' of insurance that protects your household from life's worst-case scenarios.

4.1 The Math of Dragging You Backward: Credit Mechanics

On paper, a credit card looks like a convenience. In real life, it can behave like a small leak in the hull of your household budget—quiet at first, then suddenly urgent. The leak isn't just the balance; it's the physics of interest. When you owe money at a high rate, the account becomes a machine that converts your future paychecks into yesterday's purchases. That's the headwind: you can pedal harder and still go slower, because the wind is stealing the work before it turns into progress.

The trick is that the numbers are designed to feel normal. "Minimum payment due: $___" shows up like a polite suggestion, not a trapdoor. The statement doesn't announce: "If you follow this plan, you'll rent this $10,000 for years." It just gives you a small number and a due date—and your tired brain, already juggling work, kids, and groceries, is happy to accept a small number. Credit products are engineered to be emotionally easy in the moment and mathematically expensive over time.

APR: the price tag people don't feel

APR stands for Annual Percentage Rate. That sounds straightforward, but it hides the most important detail: you're almost never charged interest once per year. Credit cards typically charge interest daily, based on a daily rate derived from the APR. A 24% APR doesn't mean "24% once"—it means roughly 24% spread across the year, applied day after day to whatever balance you carry. You can think of it as a meter running in the background.

Here's the part that changes how you see everything: interest doesn't ask your permission. It doesn't wait until you feel ready. It doesn't care that you had a car repair and then a medical copay and then a school fundraiser. Interest is automatic, and because it's automatic, it is *relentless*. When households feel like they're "doing everything right" but net worth still won't budge, this is often the culprit: the plan is trying to build a staircase while interest is quietly removing steps.

Amortization: why the early months feel like a scam

Amortization is a fancy word for a simple idea: when you make a payment on a debt, part of the payment goes to interest and whatever is left goes to the principal (the amount you actually owe). In the early stages of a debt, interest tends to dominate because the balance is still large. So you pay, and pay, and the balance barely moves. That's not because you're bad at math or lacking discipline. It's because the debt is doing what it was designed to do.

Credit cards add an extra twist: they are revolving debt, which means the "finish line" is optional. A mortgage has a schedule that assumes you'll eventually be done. A credit card is happy to keep you as a long-term tenant, paying rent to stay inside the same $10,000. The minimum payment is the rent.

Minimum payments are usually calculated as a small percentage of your balance (often around 1%–3%), sometimes plus interest and fees, with a floor (for example, "$25 minimum"). That means the payment is designed to be low enough that you can say yes. But a low payment has a side effect: it can be smaller than the rate at which interest wants to grow the balance. So the payment stops the balance from exploding, but it may not meaningfully shrink it.

Imagine a common household moment: you're on autopilot. You've got two incomes, two calendars, two stress levels, and not enough time. You tell yourself, "We'll pay it down when things calm down." Then life keeps being life. The minimum payment makes it possible to delay the hard decision—and the delay is profitable for the lender.

To make this concrete, take a $10,000 balance at 24% APR. That's not exotic; it's a normal credit card. If you paid a fixed $400 per month and never charged another dollar, you'd be done in a little over 2.5 years, and the interest would hurt but it wouldn't own you. If you paid around 2% of the balance (a common minimum-payment style), your payment starts around $200 and shrinks as the balance shrinks—which sounds nice until you realize shrinking payments also shrink your progress. You can end up paying for many years, and the interest becomes the main character in your budget.

The reason this matters isn't just "interest is bad." It's that *interest is a competitor*. Every extra dollar of interest is a dollar that can't become an emergency fund, can't buy back flexibility at work, can't reduce your fixed costs, can't become retirement savings. Debt doesn't merely cost money; it consumes options.

A real-world case study: when the card becomes the shock absorber

Consider Maya and Chris, a couple with two incomes and one constant problem: the surprises never stop. They aren't reckless. They're normal. A dental bill. A tire blowout. A flight for a family funeral. Each event is manageable, but the timing is brutal. The credit card becomes their shock absorber because it's fast and it doesn't require a conversation with a bank.

At first, the balance sits around $2,000. They tell themselves it's temporary. Then childcare costs rise and their grocery bill creeps up and one month they only pay the minimum. That month feels like relief. The next month, they do it again. The balance grows to $6,000, then $9,000. They're still making payments, so it doesn't feel like a crisis. But the interest rate means their payments are now partially *maintenance*, like paying to keep a leaky roof from collapsing rather than paying to replace it.

The turning point isn't when they "learn discipline." It's when they finally see the mechanics clearly enough to change the system. They set a rule: no new charges on the card while paying it down, even if it means making a temporary lifestyle cut and building a small cash buffer. That rule sounds obvious, but it's psychologically hard because it forces the household to feel the true cost of their monthly life. The card had been hiding the gap. Once the gap is visible, it can be solved—with spending changes, income changes, or both—but no longer with denial.

That's the real danger of revolving debt: it can make a structural problem feel like a string of unrelated emergencies. Interest stitches the emergencies into a long-term drag.

The ``interest rate tax" on your future

A helpful mental model is to treat high-interest debt like a private tax on your future earnings. Unlike actual taxes, this one isn't funding roads or schools. It's funding the privilege of having bought something earlier than you could afford it.

If you carry a balance at 24% APR, the hurdle rate is enormous. Investing in a diversified stock market index might, over long periods, average something like 7%–10% nominal returns (with lots of volatility and no guarantees). Paying off a 24% APR balance is a *risk-free* return of 24% in the sense that every dollar of balance you eliminate is a dollar that will never be charged interest again. When households try to build wealth while carrying high-interest consumer debt, it's like planting a garden while someone is mowing it nightly.

Minimum payments versus a real plan (with actual numbers)

The minimum payment is not a plan; it's a permission slip to stay in debt. A plan is a fixed monthly amount that is meaningfully larger than the minimum, paired with a behavior lock that prevents new charges from undoing the work.

Below is a visualization of what happens to the same $10,000 balance under two different approaches:

- *Minimum-payment style:* pay 2% of the current balance each month (a simplified stand-in for many issuer formulas).
- *Fixed-payment plan:* pay $400 per month until the balance is gone.

Both assume a 24% APR and no new charges.

Even if you never memorize a formula, the shape of those lines teaches the lesson: under a minimum-payment style approach, the balance doesn't just decline slowly; it declines in a way that keeps you paying interest for a long time. Under a fixed-payment plan, the balance falls with purpose. The point isn't that $400 is magic. The point is that a *deliberately chosen* payment creates an end date, and end dates change behavior.

The credit score game: what it is (and what it isn't)

Credit scores are sold as a moral grade. They're not. A credit score is a risk score: a lender's estimate of how likely you are to repay based on patterns in your credit report. It's closer to an insurance premium quote than a character judgment. That reframing is freeing, because it stops the shame spiral. Shame makes people hide statements in drawers. Curiosity makes people open them.

A credit report is the raw data file: accounts, balances, limits, payment history, delinquencies, collections, inquiries, and so on. A credit score is a summary number produced by a scoring model based on that report. There isn't only one score. Different lenders use different models and versions, and the score you see in an app may not match the score a lender pulls. This is why people sometimes swear their score "dropped for no reason." Often, the reason is simply that they're looking at one scoring system while a lender is using another. It's messy, but not mystical.

What matters for a household plan is the small set of levers that predictably move scores:

Payment history: the heavy weight

Paying on time is the big one. Late payments are costly because they are a loud signal: if you missed one payment, the model assumes you might miss others. If you're juggling cash flow, the most credit-score-efficient move is boring: automate minimum payments so you never accidentally go late, then attack the balance with extra payments when you can. Automation isn't about laziness; it's about refusing to let a hectic week turn into a multi-year penalty.

Utilization: how close you are to the cliff

Utilization means how much of your available revolving credit you're using. If you have a $10,000 limit and a $9,000 balance, your utilization is 90%. High utilization tells the scoring system you may be under strain. Even if you pay on time, maxing out cards tends to drag scores down.

Utilization has a weird emotional twist: the same household can feel "responsible" because they always pay the minimum, while the scoring model reads their profile as risky because they're consistently near the limit. The score doesn't care that you're trying. It cares what the pattern suggests.

A practical move here is to treat utilization like a dashboard warning light. If you're routinely above 30%–40% on a card, the plan isn't just to pay it down eventually; the plan is to stop living at a level that requires borrowing month after month. Otherwise, you're trying to fix a structural cash-flow issue with a number.

Inquiries and new accounts: the "what changed?" tax

When you apply for new credit, it can cause a temporary score dip because it suggests you might be about to take on more debt. The dip is not permanent, and it's not a reason to fear shopping for a mortgage or car loan responsibly. It is, however, a reason to avoid opening a store card at checkout because someone offered 10% off a blender. That discount can be the most expensive $30 you ever saved if it nudges you into a pattern of juggling new accounts.

Age and mix: slow-cooking factors

Length of credit history and the mix of account types matter, but these are slow levers. Households can waste years obsessing over them while ignoring the big two: pay on time and get balances down. If you do the big two, the rest tends to take care of itself.

The part nobody tells you: credit score optimization can be a trap

It's possible to chase a higher score while getting poorer. People do it all the time. They keep a balance "to build credit" (you don't need to pay interest to build credit; you need on-time payments). They finance a purchase they could pay cash for because the dealer offered "0%" (sometimes real, sometimes padded elsewhere). They open accounts for points and perks while their budget is already tight.

A credit score is useful because it lowers the cost of borrowing when borrowing is strategic and contained—like a mortgage you can afford or a car you plan for. But the household goal is not a pretty score. The goal is to make your life cheaper, safer, and more flexible. A higher score is a side effect of a stable system, not the system itself.

There's also a subtle psychological hazard: if you treat the score as a report card, you can start making money decisions to impress an algorithm instead of to protect your family. The algorithm doesn't care whether you slept last night. Your family does.

A simple way to ``audit" a credit card before it audits you

When a card balance feels sticky, run a quick audit that forces the truth into daylight:

- What is the APR today (not the teaser rate you remember)?
- If we pay only what we've been paying, does the balance fall meaningfully each month?
- If one income paused for 60 days, would we still make the payments without new borrowing?

That last question matters because it reveals whether the card is being used as a bridge or as a pillar. Bridges are temporary. Pillars are structural. A household that needs revolving credit to survive a short income interruption is one surprise away from the kind of stress that wrecks sleep, work performance, and relationships.

Actionable takeaway: the two rules that change everything

Pick one card you want gone and apply two rules starting this week:

1. *Create an end date:* choose a fixed monthly payment that is comfortably above the minimum, then set it on autopay. If the number is $275 or $400 or $610, it doesn't matter. What matters is that *you* chose it and it stays fixed until the balance hits zero.

2. *Install a behavior lock:* stop new charges on that card while you're paying it down. Put it in a drawer, delete it from online checkouts, and pick one alternative (debit, cash, or a different card paid in full weekly) so your life doesn't become inconvenient enough to quit.

Do those two things and you're no longer "making payments." You're converting a vague financial stressor into a project with a finish line—and every month you cross off is a month you stop renting your own paycheck.

4.2 Snowballs vs. Avalanches: Strategic Debt Payoff

At some point, almost every household with multiple debts tries the same thing: they "spread it around." A little extra on the Visa, a little extra on the car, a little extra on the store card that keeps haunting the mailbox with coupons. It feels fair. It feels responsible. It also tends to feel like pushing a shopping cart through wet cement—a lot of effort, not much movement.

Debt payoff stops feeling heroic and starts feeling effective when you pick an order and commit to it. Not because the debts have feelings, but because you do. The win isn't just mathematical; it's cognitive. A clear sequence turns debt payoff from a daily negotiation into a script you can follow when you're tired.

Two scripts dominate personal finance: the *Avalanche* and the *Snowball*. One is the cleanest math. The other is the cleanest psychology. The surprise is that the "best" method is often the one you'll actually finish.

The real enemy: decision fatigue, not algebra

If debt were only numbers, everybody would be out of it. You can download a payoff calculator in thirty seconds. You can ask a spreadsheet to be ruthless on your behalf. But debt lives inside a household—inside calendars, moods, unexpected expenses, and the kind of Tuesday where the car won't start and somebody forgot their lunch and the babysitter texts that she's sick.

This is why payoff strategy matters. You're not choosing between two formulas; you're choosing between two ways of behaving for months or years. The best plan is the one that survives normal human life: stress, boredom, setbacks, and the occasional burst of optimism that fades as soon as the water heater dies.

So yes, we'll talk about interest rates. But first, name the friction: every open balance is a tiny open loop in your brain. Another login. Another due date. Another piece of mail that smells like dread. Paying off a debt doesn't just reduce what you owe; it removes a mental tab that's been draining your attention. That attention is worth something.

The Avalanche: the math-forward method

The Avalanche method is simple:

- Pay the minimum on every debt.

- Put every extra dollar toward the debt with the highest interest rate.
- When that debt is gone, roll its payment into the next-highest rate.

If your only goal is to minimize total interest paid, Avalanche wins. High-interest debt is the most expensive debt, so you attack the most expensive fire first. In a perfectly rational world—or a household with unlimited bandwidth and zero risk of giving up—everyone would avalanche.

The Avalanche has a quiet bonus: it trains you to respect your "hurdle rate." Every extra dollar you throw at a 28% APR card is like earning a guaranteed 28% return. That's a powerful mental reset for people who have been taught to treat debt as normal and investing as the only path to wealth. With Avalanche thinking, you start seeing interest rates as gravity: you don't argue with gravity; you design around it.

But there's a catch that rarely shows up in spreadsheets: the Avalanche can feel slow at the beginning if your highest-rate balance is also your largest. Humans don't quit plans because they're wrong; they quit plans because they don't feel like they're working.

The Snowball: the psychology-forward method

The Snowball method flips the ordering:

- Pay the minimum on every debt.
- Put every extra dollar toward the smallest balance, regardless of interest rate.

- When it's paid off, roll that payment into the next-smallest balance.

Mathematically, Snowball can cost more in interest than Avalanche, especially if you delay attacking a very high rate. And yet, in the real world, Snowball often wins because it generates something the math crowd tends to underrate: momentum.

A small paid-off account is a clean win. A closed balance is a visible change in your life. One fewer payment. One fewer creditor. One fewer email reminder. When you're in debt, the emotional soundtrack is usually static—guilt, stress, avoidance, and the vague sense that you're behind. Snowball injects a different sound: progress.

There's also a behavioral phenomenon at work that's almost embarrassingly human. Many people experience what researchers sometimes call "debt account aversion": you get disproportionate satisfaction from eliminating an account entirely, even if it's not the most cost-efficient move. In other words, you don't just want less debt; you want fewer debts. Snowball is built to exploit that preference rather than fight it.

Why ``spread it around" fails (and feels like it should work)

Households often default to "spreading extra payments around" because it feels like diversification. In investing, diversification is wise. In debt payoff, it's usually a trap, because you don't get the psychological payoff of clearing an account, and you don't get the mathematical payoff of crushing a high rate. You get neither kind of win.

More importantly, spreading money around multiplies decisions. Every month becomes a fresh debate: "How much extra should go where?" That debate is a tax on your energy, and it creates openings for self-justification. On a tired month, the debate turns into: "We'll just do the minimum; we had a lot going on." A good payoff method is partly a method for removing negotiation from your life.

A concrete case study: three debts, one household, two competing plans

Take a household with these debts:

- *Card A:* $9,000 at 27% APR (big, ugly interest rate)
- *Card B:* $1,200 at 19% APR (small balance, still expensive)
- *Personal loan:* $6,500 at 11% APR (moderate rate, fixed term)

They have $600 per month they can throw at debt above the combined minimums. Not because life is perfect, but because they finally built a leaner budget and decided debt is the emergency.

Avalanche says: attack Card A first (27%). That's rational. But Card A is $9,000. The first few months, the balance barely seems to move because interest is chewing aggressively. The household feels like they're doing a lot and getting nothing.

Snowball says: kill Card B first (the $1,200). That could be gone in a couple of months. Suddenly the household has a tangible win: one card disappears, and the payment that used to go to

Card B becomes extra ammo for the next target. The psychological boost is real. You can almost hear the air come back into the room.

Now add the real-life complication: their work is stressful. One partner travels. The other handles most of the school logistics. Their odds of "sticking with the plan" are not 100%. They are not robots. In that context, a plan that creates early wins may beat a plan that is mathematically superior but emotionally punishing.

That's the central question you should ask: *which method gives us the highest probability of finishing?* Not which method wins in a vacuum. Which method wins inside your particular household.

When Avalanche is the right call

Avalanche deserves its reputation. If any of these describe your household, it's often the better choice:

- You have a very high-interest balance (mid-20s or higher) that is meaningfully increasing your monthly stress.
- You have stable cash flow and enough emotional bandwidth to stay consistent even without quick wins.
- You're the type of person who gets motivated by optimization and hates the idea of paying "extra" interest.

Avalanche also pairs well with automation. When the extra payment is scheduled and recurring, you don't need daily willpower. The plan runs even when you're not feeling inspired.

When Snowball is the right call (which is more often than people admit)

Snowball is not a math mistake; it's a behavioral design choice. It tends to work well when:

- You're overwhelmed and need visible progress to stay engaged.
- You're juggling many small debts, each with a minimum payment that clutters your monthly budget.
- You and your partner have different money personalities, and the plan needs to create shared confidence quickly.

Snowball is also a relationship strategy. Debt payoff is rarely just a finance task; it's a recurring negotiation between two adults with different fears and habits. A quick win can turn the conversation from blame to teamwork. It's hard to overstate how valuable that is. Couples don't fail at debt payoff because they can't do subtraction; they fail because resentment grows faster than the balance shrinks.

The hybrid approach that actually works in many households

Here's the compromise that often fits real life: *Snowball for traction, Avalanche for efficiency.*

That looks like this:

- First, pay off one or two small balances to generate momentum and free up cash flow.
- Then, switch to Avalanche and attack the highest-interest remaining debt with your now-larger monthly payment.

This isn't fence-sitting; it's sequencing. Early wins reduce stress, and reduced stress increases consistency. Consistency is what makes math matter.

If the phrase "switch methods" makes you nervous, that's a sign you may be craving a single set of rules you don't have to revisit. That's fine. But recognize the tradeoff: simplicity is valuable, and so is not paying unnecessary interest. The hybrid method is for households that can handle one strategic pivot.

The non-negotiable rule either way: stop adding new debt

No payoff method survives if new debt keeps flowing in. This is where households get stuck: they start a plan, then an expense hits, and the credit card becomes the pressure release valve again. The plan collapses, and the household concludes they "can't get ahead."

It's not a character flaw; it's a systems flaw. If your budget has no margin, debt payoff becomes an optimism project. You need at least a thin layer of protection—a starter emergency fund, a sinking fund for predictable costs, or a temporary reduction in discretionary spending—so that the next minor crisis doesn't erase months of work.

A payoff method is a targeting system. But you also need a shield.

A vivid way to choose: which pain do you prefer?

Choosing between Snowball and Avalanche is often choosing between two types of pain:

- *Avalanche pain:* slower emotional rewards early on, but a cleaner long-term cost.
- *Snowball pain:* potentially higher interest cost, but faster visible wins and fewer monthly payments sooner.

Neither is wrong. The wrong move is picking a method that makes you quit.

If you've started and stopped debt payoff plans before, take that history seriously. Your past behavior is data, not a reason for shame. If you tend to burn bright for six weeks and then fade, you don't need a more aggressive spreadsheet. You need a method that keeps you in the game when motivation disappears.

How to run the plan as a household (without turning it into a monthly fight)

Debt payoff is one of those tasks that can quietly poison a relationship if it becomes a constant performance review: who spent what, who caused the balance, who is "better" with money. That dynamic is both painful and counterproductive.

A better way is to treat the plan like a shared operating system:

- One target debt at a time gets the extra money.
- Both partners can see the scoreboard (balances and payoff dates), but neither partner gets to use it as a weapon.

- Lifestyle cuts are chosen deliberately and temporarily, like turning down the thermostat in winter to get through a storm.

If one partner is more anxious, make the plan more visible. If one partner is more avoidant, make the plan more automatic. Household finance is often the art of designing around personalities rather than demanding personalities change on command.

The moment that makes Snowball powerful: the payment roll

The most satisfying part of Snowball (and also Avalanche) is the roll: when a debt is paid off, you don't "reward yourself" by spending the freed-up payment. You roll it into the next target. Your total monthly debt attack number stays the same, but the power behind it grows each time you eliminate an account.

This is how a plan goes from "we're trying" to "we're dangerous." At the start, you might only have $150 extra. After two payoffs, you might have $600 extra. Later, you can have $1,200 extra. The snowball becomes real not because of the balances, but because of the *payment velocity* you build.

If you've never experienced this, it's hard to believe. People assume debt payoff is linear: pay, pay, pay, slowly improve. In practice, it can be lumpy. The first payoff takes effort. The next one is faster. Later, balances can fall with startling speed. That acceleration is the emotional fuel that keeps households going.

A practical, immediate takeaway: choose your method in 15 minutes

Set a timer for 15 minutes and do this on paper (or a shared note):

1. Write every non-mortgage debt balance, minimum payment, and interest rate.
2. Circle the smallest balance and the highest interest rate.
3. Answer one question honestly: *Do we need quick wins to stay motivated?*

If the answer is yes, start with the Snowball: smallest balance first. If the answer is no—or if the highest-rate debt is so expensive it feels like it's eating the budget alive—start with Avalanche: highest rate first.

Then make it real: schedule the extra payment as an automatic transfer for the day after payday, and put the target debt somewhere you can see it (a sticky note, a tracker on the fridge, a line in your budgeting app). The point isn't to become a person who loves budgeting. The point is to build a system that works even when you're not at your best.

Finish the day by doing one small, concrete thing that reduces the odds of backsliding: delete saved card numbers from shopping sites, freeze a card (literally, in a bag of ice if you have to), or move one subscription to a cheaper plan. A debt payoff method is important. A debt payoff *environment* is what makes the method stick.

4.3 The Mortgage You Can't Live In: Managing Student Loans

Student loans are the only "mortgage" that doesn't come with a roof. You can't move into it, rent out a bedroom, or sell it when you're tired of it. It just sits there in your financial life like a second gravity field, pulling on your cash flow month after month—often while you're also trying to pay for child-care, save for retirement, and keep your car from turning into a surprise fundraiser.

The emotional punch comes from how student debt *feels* compared to other debt. A credit card balance usually carries a sharp edge: you can see the interest. A student loan can feel oddly polite, especially if the payment is fixed and the interest rate isn't outrageous. The danger is that "polite" debt can quietly block everything else. The household keeps functioning, but the future gets postponed in small, expensive ways: less flexibility at work, smaller emergency fund, delayed investing, and the constant low-level stress of owing a lot for a long time.

A useful goal isn't "pay it off as fast as possible." The useful goal is: *manage student loans so they don't freeze the rest of your financial life*. That means balancing math, risk, and policy reality.

Start by naming what kind of debt you have (because the rules are wildly different)

Student loans come in two major flavors: federal and private. They look similar on a budget line, but they behave differently when life gets weird.

Federal loans are governed by federal rules: income-driven repayment options, potential forgiveness programs, and protections like deferment/forbearance (with details that matter). Private loans are basically standard consumer loans: fewer escape hatches, fewer forgiveness paths, and less flexibility if income drops. In exchange, private loans can sometimes offer lower interest rates if you refinance—but refinancing federal loans into private loans is a one-way door. Once you walk through it, the federal options don't come with you.

So the first move is an inventory, not an emotion:

- Who is the servicer?
- Is each loan federal or private?
- What is the interest rate on each loan?
- What repayment plan are you on right now?
- Is either spouse pursuing Public Service Loan Forgiveness (PSLF) or any employer-based assistance?

If you can't answer those in a single sitting, that's not a personal failing. It's how the system is designed: complicated enough that avoidance feels reasonable. But clarity is the difference between "we have debt" and "we have a plan."

A case study: the household that tried to outrun the payment

Janelle is a nurse, Marco works in tech support, and they have two kids. Their combined student loans are $96,000—a mix of federal loans from Janelle's nursing program and Marco's older private loan. When daycare was at its peak, their household strategy was simple: survive. They paid what was required and tried not to look too closely.

Then their oldest started kindergarten and daycare costs dropped. They felt a surge of optimism and decided to "finally crush the loans." The plan was aggressive: throw every freed-up dollar at the balance. For three months, it worked. Then the car needed a major repair, then Marco's hours got cut for six weeks, and suddenly the aggressive plan turned into late nights, stress, and a creeping sense of failure.

The turnaround wasn't about willpower. It was about redesign.

They separated the problem into two tracks:

- Stabilize the household: a starter emergency fund and a monthly "predictable surprises" sinking fund (car repairs, school expenses, medical copays).
- Optimize the loans: choose a sustainable payment for federal loans and refinance the private loan only if the numbers clearly improved.

That second track mattered, but the first track made it possible. Without a cash buffer, every aggressive debt plan is one broken alternator away from collapse.

The three-job framework: cash flow, interest, and optionality

Student loans demand a plan that does three jobs at once.

Job 1: Protect cash flow. Your monthly payment must fit real life, not fantasy life. If a payment forces you to run up credit card debt, it's not a "responsible" payment. It's a payment that creates more expensive debt.

Job 2: Respect interest rates. High-interest balances are a headwind. A 7% student loan isn't a credit card, but it's still costly. The higher the rate, the more attractive aggressive payoff becomes.

Job 3: Preserve optionality. Federal student loans come with options that can be valuable when income changes, when a borrower becomes disabled, or when forgiveness programs apply. Those options have a real dollar value, even though you can't see them on a statement.

Households get into trouble when they optimize one job and ignore the others. For example: paying aggressively (interest job) while running a thin emergency fund (cash flow job) often backfires and pushes the household toward credit cards. Or refinancing federal loans (interest job) without valuing forgiveness/protections (optionality job) can become a permanent regret if a job change later would have qualified for PSLF.

Income-driven repayment (IDR): the tool that is helpful and messy

Income-driven repayment plans are designed to set payments based on income and family size, not just what you owe. In theory, they can make payments manageable and provide forgiveness after a long period of qualifying payments.

In practice, IDR can feel like trying to hit a moving target while the rules are being rewritten. It's not paranoia; the last few years have been unusually volatile. For example, on March 26,

2025, the U.S. Department of Education announced it had reopened the online IDR and loan consolidation applications after a temporary pause tied to a court injunction affecting SAVE and parts of other IDR plans. Then, on December 9, 2025, the Department announced an agreement framework (pending court approval) that would end SAVE: no new enrollments, denial of pending SAVE applications, and moving existing SAVE borrowers into other plans.

Even if you never used SAVE, the lesson matters: *student loan policy can change on timelines that don't respect your household calendar*. A plan that relies on a single program as if it's permanent is fragile. A plan that assumes change and keeps documentation is resilient.

That doesn't mean "don't use IDR." It means use it like a tool, not like a promise carved into stone.

The two classic mistakes: paying too little forever, or paying too much too soon

Student loans create two equal-and-opposite traps.

Trap A: The forever minimum. The payment is manageable, so the household never revisits it. Years pass. Income rises. The payment stays low. The balance barely moves. Then the borrower wakes up one day with a larger income, a still-huge balance, and the creeping dread that they have "wasted" their thirties.

Trap B: The heroic sprint. The household tries to annihilate the balance immediately, starving everything else: emergency fund, retirement contributions, home maintenance, mental health. This works for a short time, then a normal life event hits, and the plan breaks.

The right approach for many households is neither complacency nor martyrdom. It's staged intensity: stabilize first, then accelerate.

Refinancing: when it's brilliant and when it's a trapdoor

Refinancing student loans means taking a new loan (usually private) with a different interest rate and term. Refinancing can be a powerful move for private loans and, in some cases, for federal loans *if you are absolutely sure you don't need federal protections or forgiveness options*.

Refinancing is often brilliant when:

- The loan is private already, with a high rate.
- Your credit and income have improved since you took it out.
- The new rate meaningfully lowers interest cost without stretching the term so long that you pay forever.

Refinancing is often a trapdoor when:

- The loans are federal and you may be eligible for PSLF.
- Your household income is variable or job security is uncertain.
- You're leaning on federal options like income-driven repayment, deferment/forbearance pathways, or potential discharge provisions.

A simple household test: if you refinanced federal loans into private loans and then lost one income for three months, would you still be okay? If the answer is "we'd be in trouble," the federal safety net may be worth more than the interest savings.

PSLF: the program that rewards career choices (but punishes sloppy paperwork)

Public Service Loan Forgiveness can be life-changing for borrowers working in qualifying public service jobs. The big idea is straightforward: make qualifying payments on a qualifying plan while working for a qualifying employer, and the remaining balance may be forgiven after the required number of payments.

The details are where households get hurt. PSLF is less like a coupon and more like a punch card that only works if each punch is stamped correctly. That means employment certification, correct loan types, correct repayment plan, and accurate tracking. A borrower can do the work for years and still stumble if they don't document.

If PSLF is on the table in your household, the plan should include a recurring calendar event for paperwork and verification. Make it boring, make it routine, make it non-optional. No one should be trying to reconstruct ten years of details from memory.

The ``parallel track" strategy: keep life moving while the loans get handled

Households often treat student loans like a moral debt: "We can't invest until we pay this off." That sounds noble. It can also be expensive, especially when the loans are moderate-interest federal loans and the household is losing years of retirement compounding.

A more useful approach is a parallel track:

- Make a sustainable required payment on student loans (or a strategically chosen plan payment).
- Simultaneously build a small emergency fund and contribute at least enough to retirement to capture any employer match.
- Add extra principal payments only when the household has margin (and only if those extra payments don't force you back to revolving debt later).

This keeps the household from living in a financial holding pattern. Student loans are long. Your life is happening now.

Policy volatility: plan for the world as it is, not as it should be

Student loans are unusual because the "rules" are partially political and partially legal. That's not a reason to obsess over headlines; it's a reason to build a plan that doesn't shatter when headlines change.

Three practical ways to do that:

Keep your own records. Download key documents, payment histories, and confirmation numbers. If a portal is down, if a servicer changes, if processing backlogs grow, you still have proof of what you did and when.

Avoid cliff-edge budgeting. If your plan only works when the payment is exactly $0 or exactly a particular low number, you've built a fragile system. Aim for a household budget that could survive a payment increase. Not happily. Not forever. But survivably.

Treat your loan plan like a setting, not an identity. You can be the kind of person who pays debts responsibly without swearing allegiance to one repayment program. Flexibility is not weakness; it's competence.

How to decide what to do with extra money (without getting stuck in analysis)

When a household has extra cash each month, student loans compete with other good uses: emergency fund, retirement, childcare, home repairs, and sometimes medical needs. The way out of paralysis is to use a simple priority ladder:

First, eliminate any behavior that creates new high-interest debt (because that undoes every other choice). Next, build enough cash buffer that a normal surprise doesn't send you back to the credit card. Then, decide where student loans sit relative to retirement saving based on interest rate and forgiveness likelihood.

If you're staring at a 6.8% federal loan and you're not pursuing PSLF, extra payments can be a strong move once the household is stable. If you're on a path where forgiveness is likely and the payment is already optimized, overpaying may be less urgent than building assets elsewhere. The "right" answer isn't universal; it depends on which future you're buying.

Actionable takeaway: a 30-day student loan reset that doesn't hijack your life

Over the next 30 days, do these in order:

1. *Build the one-page inventory.* List every loan, whether it's federal or private, the interest rate, servicer, balance, and required payment. If you're married, put both spouses' loans on one sheet. Seeing the whole picture is calming in a way people don't expect.
2. *Choose one "stability buffer" number.* It can be $500, $1,000, or one month of bare-bones expenses. The number isn't the point; the point is to stop using debt as your emergency fund.
3. *Set one automatic payment and one automatic "extra."* Automate the required payments, and automate a small extra payment toward the chosen target loan (even $25 or $50). This builds the habit and creates forward motion without demanding perfection.
4. *Create a policy-proof folder.* Download your loan details, recent payment history, and any plan confirmations. Put them in one place. Future-you will feel like someone finally has your back.

At the end of that month, you won't necessarily owe dramatically less. You'll have something better: a household system that can handle student loans without putting the rest of your life in a waiting room.

4.4 The Risk Matrix: What to Insure vs. What to Budget

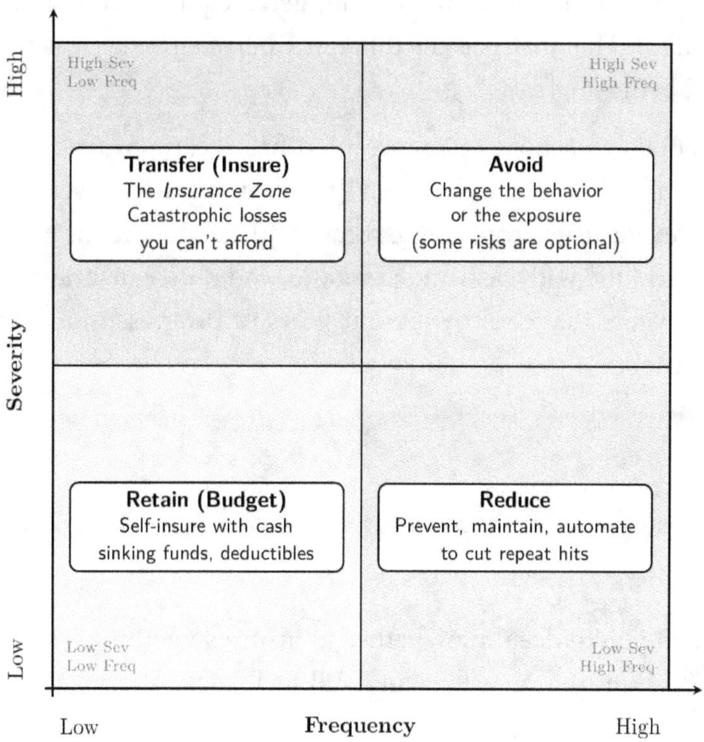

Insurance is one of the few products you buy hoping you never get your money's worth. You hand over real dollars for a promise you pray stays hypothetical. And yet most households still manage to overpay for the wrong things and underprotect the stuff that could actually ruin them. It's not because they're careless. It's because the world of risk is noisy: every commercial is a fear button, every neighbor has a story, and every policy is written in a dialect that seems designed to make normal people give up and sign.

The way out is not to memorize every coverage type. The way out is to sort risks like an engineer sorts problems: by how often they happen and how bad they get. Frequency and severity. That's it. That matrix is the difference between being "insured" and being *protected*.

A good household insurance plan has a certain personality. It's calm. It's boring. It doesn't try to prevent inconvenience; it prevents catastrophe. It assumes you can handle the annoyances of life with cash and planning, and it uses insurance for the events that could wipe out years of progress in a single afternoon.

Frequency and severity: the two questions that beat 90% of confusion

When you're deciding whether to insure something, you can ask a hundred questions and still feel stuck. Or you can ask two:

- *How likely is it?* (Frequency)
- *If it happens, how expensive is it?* (Severity)

Most household financial pain comes from mixing these up. People buy insurance for frequent, small problems because it feels comforting. Then they skip insurance for rare, devastating problems because it feels "unlikely." But the whole point of insurance is that it's designed for low-frequency, high-severity events: the kind of loss you can't just "tighten the belt" and absorb.

This is also why deductibles exist. A deductible is the portion of a claim you agree to pay yourself. It's a built-in boundary between what you budget for and what you transfer to an insurance company. Set that boundary thoughtfully and your premiums often drop. Set it thoughtlessly and you'll pay high premiums to insure the kind of losses you could have handled with a modest emergency fund.

Quadrant 1: Retain (Budget) --- low frequency, low severity

This quadrant is the land of normal life. A cracked phone screen. A small appliance dying. A tire replacement. The kind of expenses that are irritating but not life-changing.

If you insure every small, manageable event, two things happen:

- You pay for the insurer's overhead and profit (which is real).
- You keep yourself dependent on claims and reimbursements, which is slow, paperwork-heavy, and often stressful.

Retaining the risk doesn't mean pretending bad things won't happen. It means pre-paying yourself through sinking funds:

little buckets of money reserved for predictable categories. A "car repairs" fund isn't pessimism; it's realism with better manners. When the inevitable happens, you're annoyed, not financially destabilized.

The emotional win is bigger than the math. When your household has cash for the annoying stuff, you stop treating every inconvenience like a crisis. And when you stop living in crisis mode, you make better decisions everywhere else.

Quadrant 2: Reduce --- high frequency, low severity

This is where habits beat policies.

If something happens often but doesn't cost much each time, insurance is usually a clumsy tool. The smarter move is to reduce frequency through prevention and systems. This is the quadrant of smoke alarms, routine maintenance, and boring discipline that saves real money.

A few examples that belong here:

- Raising your auto deductible (which lowers premium) *only* after you've built a cash buffer that can cover it.
- Home maintenance that prevents repeat damage: gutter cleaning, small plumbing fixes, servicing HVAC before it fails at the worst possible moment.
- Cyber hygiene: password manager, two-factor authentication, freezing credit when appropriate.

This quadrant is a reminder that not all "risk management" is insurance. Sometimes it's a calendar reminder and a trip to the hardware store.

Quadrant 3: Transfer (Insure) --- low frequency, high severity (the Insurance Zone)

This is the money-saving heart of the matrix: the events that are unlikely but financially lethal.

Think: house fire, major liability lawsuit, a serious car accident with injuries, a medical event that triggers long work absence, theft, a storm that destroys part of your home. If these hit and you're underinsured, the problem isn't the bill. The problem is that the bill changes the entire trajectory of your life.

Insurance belongs here because it turns a life-altering loss into a manageable loss. You're buying a ceiling on the damage.

A hard truth: many households accidentally insure the wrong thing in this quadrant by focusing on *property* while neglecting *liability*. Replacing a couch is painful. Defending a lawsuit or paying a judgment can be existential. Liability is also where the numbers get surreal quickly, because it's not just your assets at stake; it can be future wages and future savings.

This is also where the deductible conversation becomes powerful. A higher deductible is a vote of confidence in your emergency fund. If you can comfortably cover a $1,000 or $2,500 deductible, you can often reduce your premium and stop paying the insurer to handle medium-sized inconveniences. But if you set a high deductible without the cash to back it up, you've created a paper plan. The first claim becomes a financial scramble.

Quadrant 4: Avoid --- high frequency, high severity

Some risks are both likely and devastating. That's the quadrant that should make you change behavior, not shop for coverage.

Driving without adequate auto liability limits? That's avoid. Letting an unlicensed teen drive your car? Avoid. Operating a side business without understanding whether your homeowners policy excludes business-related liability? Avoid. Keeping an aggressive dog breed without appropriate coverage in a state or insurer environment that treats it harshly? Avoid.

This quadrant is not about being fearful. It's about refusing to play games with outcomes that can wreck your household. Avoidance can look like changing routines, changing exposures, or sometimes changing what you own.

A real-world case study: the deductible that broke the plan

A family—Sam and Leila, two incomes, one child—decided to "be responsible" and bought rich coverage on everything. Low auto deductible, low homeowners deductible, extra riders for minor stuff. Their premiums were high, but they felt safe.

Then a small kitchen leak caused damage. The repair cost wasn't catastrophic, but it was annoying. They filed a claim because, after all, that's what insurance is for. The claim was handled, but later their renewal premium jumped. They were surprised and angry. They did what many people do next: they raised deductibles to lower premiums.

The problem: they raised deductibles without building cash reserves.

A year later, a hailstorm damaged the roof. This time, the loss was legitimately large and clearly belonged in the Insurance Zone. The claim was approved. But the deductible was now $2,500, and they didn't have $2,500 sitting around. They put the deductible on a credit card, then carried the balance. They had swapped one risk for another: they reduced premium but increased the chance that a claim would create high-interest debt.

Their fix was beautifully unglamorous:

- They set deductibles at a level they *could* cover.
- They built a home-maintenance sinking fund so small claims didn't become insurer claims.
- They reserved insurance for the big losses and stopped treating it like a reimbursement app.

The punchline isn't "never file a claim." The punchline is that your insurance strategy and your cash strategy are the same strategy. If they aren't coordinated, the first bad day exposes the gap.

Homeowners insurance: what you're really buying

Most people think homeowners insurance is about replacing the building. That's part of it, but a typical homeowners policy is more like a bundle:

- *Dwelling coverage* (the structure)
- *Personal property* (your stuff)

- *Loss of use* (extra living expenses if you can't live there after a covered event)
- *Personal liability* (if someone is hurt or property is damaged and you're responsible)
- *Medical payments* (small payments for injuries on your property, depending on policy terms)

Two household surprises show up again and again:

First, flood and earthquake are often *not* covered by standard homeowners policies. If you live in an area with meaningful flood risk, "we have homeowners insurance" can be an expensive misunderstanding.

Second, the difference between replacement cost and actual cash value matters. Replacement cost aims to pay what it costs to replace something today. Actual cash value subtracts depreciation, which can turn a claim into an unpleasant math lesson. When a catastrophic loss hits, depreciation is not the moment you want to discover you insured your life at a discount.

Auto insurance: the quiet liability monster

Auto insurance is where many households accidentally take on huge risk to save a few dollars per month. The danger isn't the car; it's the people. Injuries can be expensive, and fault can be complicated. If your liability limits are low and an accident is severe, you can be exposed in ways that ripple for years.

A practical way to use the matrix:

- Fender-benders and minor repairs are frequent-ish and usually manageable with a deductible and a sinking fund (retain/reduce).
- Serious accidents with injuries are low frequency, high severity (transfer/insure).

That usually pushes households toward higher liability limits, while using deductibles strategically.

The credit-score-to-insurance-cost connection that surprises people

Many households treat credit as a borrowing problem only. But in some states and with some insurers, credit report information can affect insurance pricing through what's often called a credit-based insurance score. That means sloppy credit management can show up not only as higher loan interest, but as higher premiums.

This is not an argument to obsess over a number. It's an argument to see your financial system as interconnected. When you pay bills on time, keep utilization reasonable, and avoid unnecessary new credit, you're not just making lenders happier. You may be making insurance cheaper too, depending on where you live and which company underwrites your coverage.

How to choose deductibles like an adult (not like a gambler)

Deductibles are one of the most powerful levers you control. They are also where households most often make performative choices: setting a deductible that looks tough on paper but collapses at the first claim.

A clean rule of thumb: choose deductibles that you can pay *tomorrow* without borrowing.

Not "we could figure it out." Not "we'd put it on a card for a minute." Tomorrow, without debt.

Then run the quote test: get premium quotes at different deductible levels and compare. When you see how much you save by raising a deductible from $500 to $1,000 (or $1,000 to $2,500), you can make a deliberate trade: "We'll self-insure the first $X, and we'll pay less every month for the catastrophic protection." That's the matrix in action.

Actionable takeaway: build your household risk plan in one sitting

Set aside 45 minutes and do this with your partner (or by yourself if you're the household CFO):

1. Draw the frequency/severity matrix on paper.
2. Write ten risks your household actually faces (not theoretical ones): car accident, job loss, roof damage, medical event, phone breakage, plumbing leak, lawsuit risk, pet bite, identity theft, travel disruption, etc.
3. Place each risk in a quadrant and write the matching action beside it: Avoid, Retain (Budget), Reduce, or Transfer (Insure).
4. For every "Transfer (Insure)" risk, check whether you truly have coverage and whether the deductible is payable tomorrow without debt.

When you're done, the goal isn't to feel perfectly safe. The goal is to feel *appropriately* safe—protected from the disasters, prepared for the annoyances, and no longer paying insurance premiums to solve problems your emergency fund could handle more gracefully.

4.5 Insuring the Golden Goose: Income Protection

If someone stole your car, you'd feel violated. If someone stole your laptop, you'd feel exposed. If someone stole your ability to earn an income, you wouldn't just feel panicked—you'd be standing in the middle of your life with the power shut off.

That's the uncomfortable truth: for most households, the most valuable asset isn't the house or the retirement account. It's the paychecks you expect to collect over the next twenty years. Two-income households feel especially stable because the bills are spread across two earners. But that stability can be a mirage if the plan assumes both incomes will always show up on schedule. Life doesn't sign that contract.

Income protection is the quiet kind of grown-up planning. It's not exciting. It doesn't come with influencer screenshots or shiny bragging rights. It's the financial equivalent of installing seatbelts in a car you hope never crashes. And when the crash happens, it's the difference between a hard season and a permanent detour.

A case study: the year the household became a one-income household overnight

Tanya and Eric had what looked like a solid setup: two incomes, one mortgage, two kids, and a decent rhythm. Their budget wasn't perfect, but it worked. Tanya carried the health insurance through her job; Eric brought in a slightly higher salary and most of the retirement contributions.

Then Eric hurt his back lifting something ordinary—not a dramatic workplace accident, just the kind of injury that happens in real life. At first it seemed like a "couple of weeks" situation. Then the pain didn't resolve. Then physical therapy. Then imaging. Then a doctor told him to stop working for a while.

Here's what surprised them: the problem wasn't only the medical bills. The problem was time.

They had some sick leave. Then short-term disability through Eric's employer kicked in, but it didn't replace his full paycheck. Their fixed costs didn't care. The mortgage didn't shrink because Eric was in pain. Daycare didn't offer a discount because they were stressed. Groceries didn't get cheaper because they were scared.

They assumed that if things got truly bad, there was "disability from the government." That's when they found the first trapdoor: Social Security Disability Insurance (SSDI) is not a quick rescue. Even for people who qualify, SSDI has a waiting period: benefits generally don't begin until the sixth full month after disability onset (a five-month waiting period). And even after SSDI begins, Medicare typically starts after 24 months of SSDI entitlement—a timeline that can leave households exposed if they lose employer-sponsored health coverage and don't have a backup plan.

What saved them wasn't heroics. It was that Eric had elected employer long-term disability coverage years earlier, almost as an afterthought, because the premium felt small. That policy became the financial bridge that kept them housed while Eric recovered. Without it, they would have drained retirement, run up credit cards, and possibly sold the house at the worst moment.

The point isn't that every injury becomes a long-term disability. The point is that income loss is often a *cash-flow event first*. The household needs time to breathe. Disability insurance buys time.

The three ways income disappears (and why most people only plan for one)

Most households plan for job loss, sort of. They know layoffs happen. They have a vague mental model: "We'd cut expenses and find something else." That's not a plan, but at least it's a category.

Income can also disappear through two other doors that don't get enough attention:

Disability or serious illness. This includes injuries, chronic conditions, cancer treatment, mental health episodes, complications from surgery, autoimmune flare-ups—real life. Sometimes you can work partially. Sometimes you can't. Sometimes the recovery is not linear.

Death. It's blunt to say, but it's clean math: if an income stops permanently because a person is gone, the bills don't politely stop with them.

Income protection means addressing all three with the right tools: emergency funds for job loss and short gaps; disability insurance for disability; life insurance for death. The mistake is treating them as interchangeable.

Disability insurance: the most ignored ``must-have" in many households

Disability insurance replaces a portion of your income if you can't work due to illness or injury. If you're thinking, "I have health insurance," pause. Health insurance pays doctors and hospitals. It does not replace your paycheck.

Disability insurance is a paycheck policy.

That framing matters because it changes how you evaluate it. You're not shopping for a product; you're protecting the machine that pays for your entire life. Two-income households often underestimate this because the second income creates a sense of redundancy. But redundancy isn't the same as resilience. If one income disappears, the remaining income may not cover the full system. And even if it can, the strain can be severe: one person becomes the sole provider while also carrying emotional and logistical weight.

Short-term vs. long-term disability: the gap is where households drown

Many employers offer two layers:

Short-term disability (*STD*) often covers the first weeks or months of a disability. It usually replaces a percentage of income (not 100%), and it may require you to use sick days first.

Long-term disability (*LTD*) covers longer periods, potentially years, depending on the policy. LTD is the one that protects the household from the slow-motion financial collapse that happens when a medical event becomes a long recovery.

The elimination period (also called the waiting period) is crucial. It's the time you must be disabled before benefits begin. Common elimination periods include 30, 60, 90, or 180 days. Choosing a longer elimination period typically lowers premiums, but it increases the amount of cash you must have to bridge the gap. This is where the household system becomes either elegant or dangerous: if you have a strong emergency fund, you can "self-insure" a longer waiting period. If you don't, a long elimination period is just a polite way of saying "good luck."

The definition of disability: the fine print that matters more than the premium

Disability insurance policies are full of definitions, because definitions control payouts. The big one is the definition of "disabled." Two common concepts show up:

Own-occupation (often shortened to "own-occ") generally means you're considered disabled if you can't perform the duties of your specific occupation.

Any-occupation means you're considered disabled only if you can't perform *any* job you're reasonably suited for.

Those sound similar until you imagine the difference. A surgeon with hand tremors might be unable to operate but could technically teach. Under a strict any-occupation definition, they might not qualify. Under an own-occupation definition, they might.

Not everyone needs the most premium version of every feature. But households should treat definitions like load-bearing beams. Paying a little less for coverage that doesn't pay when you need it is not savings; it's self-deception.

Many disability policies also include limitations for certain conditions (for example, mental/nervous conditions can have different benefit periods). Again: not to scare you, but to encourage reading the outline of coverage and asking questions before you're stressed and ill. Good consumer standards around disability products emphasize providing a clear outline of coverage—and you should insist on it.

SSDI: a backstop, not a household plan

It's reasonable to ask, "Isn't there government disability?" Yes, SSDI exists. But it's a backstop, not a first-line plan for a working household.

Two timelines are especially important:

- SSDI benefits generally do not begin until the sixth full month after disability onset (a five-month waiting period).

- Medicare generally begins after 24 months of SSDI entitlement.

Even if a household ultimately qualifies, those gaps matter. They are exactly the kind of gaps that push people into high-interest debt, early retirement withdrawals, or desperate decisions. A good household plan assumes SSDI is uncertain, slow, and partial. If it arrives, it helps. But you don't build your entire roof expecting a parachute to deploy.

Life insurance: cheap, boring, and massively important when someone depends on you

Life insurance is for the people who would be financially harmed by your death. That could be a spouse, kids, aging parents, or anyone who relies on your income, childcare labor, or benefits (like employer-sponsored health insurance).

The most efficient product for most working families is term life insurance: coverage for a specific term (often 10, 20, or 30 years). It's designed for the season when your income is most necessary and your obligations are highest. The reason term tends to fit is simple: your need for life insurance usually declines over time as savings grow, the mortgage shrinks, and children become independent.

Permanent life insurance (whole life, universal life, and variations) combines insurance with an investment component and typically costs far more. There are situations where permanent insurance can make sense, but for the typical two-income household trying to escape financial fragility, it often becomes an expensive detour. Paying high premiums for permanent policies while carrying consumer debt or underfunding retirement is like buying a luxury safe while the front door is still unlocked.

How much life insurance is "enough"?

The point is not to perfectly replace every future dollar. The point is to prevent a tragedy from becoming a financial collapse.

A practical way to think about it:

- What debts would you want eliminated immediately? (Often the mortgage and any consumer debt.)
- What income needs to be replaced, and for how long? (Often until kids are grown or until the surviving spouse can adjust work and childcare.)
- What non-income labor is being replaced? (Childcare is not free just because an income disappears; it often gets *more* expensive.)

Two-income households sometimes underinsure because they assume the surviving spouse will "just work more." That can be unrealistic when grief, parenting, and logistics collide. Insurance is a way of buying the surviving spouse options: time off, reduced hours, childcare help, the ability to stay in the home, the ability to avoid panic decisions.

The hidden income-protection issue: benefits that disappear with the job

Income protection is not just salary. It's also the scaffolding attached to employment: health insurance, employer retirement match, and sometimes life and disability coverage provided at work.

This creates a planning blind spot: households assume benefits are stable until the day they aren't. If one spouse carries health insurance and becomes unable to work, the household may face not only income loss but a health coverage scramble. That's why an emergency fund is part of income protection even when you have insurance products. Money buys time to make careful decisions instead of frantic ones.

A household checklist that avoids the common traps

You don't need to become an insurance nerd. You need to ask the questions that keep you from being surprised in the worst week of your life.

For disability coverage (especially employer coverage), find:

- The benefit percentage (e.g., 60% of income) and whether there's a cap.
- The elimination period (how long until benefits begin).
- The benefit period (how long benefits can last).
- The definition of disability (own-occupation vs. any-occupation, and how it changes over time).
- Any major limitations or exclusions that would matter for your situation.

For life insurance, confirm:

- The term length lines up with your highest-obligation years.
- The beneficiary designations are correct (and updated after life changes).
- The coverage amount actually matches the household's dependency reality, not just a guess made at age 27.

Then match your deductibles and waiting periods to cash. If you choose a 90-day elimination period for LTD and your household has two weeks of savings, you didn't build a plan; you built a hope.

Actionable takeaway: the ``Friday night" income protection audit

Pick one evening this week and do a one-hour audit. Not a research rabbit hole, not a shopping spree—just an audit.

1. Pull the benefits summaries from both employers (disability and life, if offered) and write down: benefit percentage, caps, waiting period, benefit period.
2. Write down the household minimum monthly expenses (housing, utilities, groceries, insurance, debt minimums, childcare).
3. Do one uncomfortable calculation: if one income stopped tomorrow, how many months could you keep paying bills *without* borrowing?

4. If the answer is "not long," choose one next step you can actually execute: enroll in employer LTD during open enrollment, price term life insurance, or build the cash buffer that makes your waiting periods survivable.

The goal is not to eliminate risk. The goal is to make sure the most valuable thing your household owns—the ability to earn—is protected well enough that a bad diagnosis, a bad accident, or a bad day doesn't get to rewrite the rest of your life.

4.6 Building the Forcefield: Liability and Umbrella Coverage

Most people think the financial dangers are obvious: job loss, medical bills, maybe a market crash. Lawsuits feel like a problem for celebrities and corporations—people with yachts and headlines. Then you hear a story that rewires your brain: a normal car accident, a routine guest injury, a split-second mistake, and suddenly the numbers on the page don't look like "a bill." They look like *a second life you now have to pay for*.

Liability risk is the risk that you harm someone else (or damage their property) and you're legally responsible for the cost. It's the kind of risk that feels abstract until it isn't. And it's uniquely threatening to a two-income household because it targets the very thing you're trying to build: accumulated savings *and* future income. You don't need to be reckless to be exposed. You only need to be present in the modern world: driving, owning property, having a dog, hosting friends, employing a babysitter, coaching a team, posting online, renting cars on vacation.

If debt was the headwind, liability is the lightning strike. Rare, violent, and capable of undoing years of careful progress in one event. The goal here isn't to become paranoid. The goal is to build a forcefield: high enough limits, the right structure, and a plan you can live with.

The most misunderstood sentence in insurance: ``I'm fully covered"

People say "we're fully covered" the way they say "we have groceries." It's a comforting phrase, not a precise statement. Insurance doesn't work like that. You're always covered up to a number, under specific rules, with exclusions and conditions. Liability coverage is not a magic shield; it's a contract.

The two contracts most households already have are:

- *Auto insurance* (liability for injuries and property damage you cause in a vehicle-related incident)
- *Homeowners or renters insurance* (personal liability, plus property coverage and other pieces)

The part that matters for forcefield-building is the liability limit—the maximum the insurer will pay for covered liability claims. If your limits are low and the damages are high, the difference is on you. Your savings. Your paycheck. Your future.

That's why liability coverage isn't about replacing stuff. It's about defending your life's balance sheet.

A case study: the accident that turned into a math problem

Miguel and Priya were doing well by normal standards: two incomes, a growing 401(k), and a house they bought before prices went completely feral. Their auto policy had the same liability limits they'd carried since their twenties, because they'd never thought to change it. The premium was low. The app said they were "good."

Then Miguel rear-ended a car at highway speed in bad weather. The other driver had serious injuries. No one died. No one was being reckless. It was just a bad second.

The medical bills, lost wages, and pain-and-suffering claim added up quickly. The number was far above Miguel and Priya's auto liability limit. Their insurer paid up to the limit, and then the negotiation got personal. They were suddenly learning phrases they'd never wanted to know: "policy limits," "excess exposure," "subrogation," "judgment."

They didn't lose everything, but they did lose something that's harder to quantify: peace. They spent months in a low-grade panic, cutting spending and delaying plans, not because they wanted to save money, but because they were trying to build a defensive pile in case the worst happened.

After it was resolved, they made a decision that would have seemed dramatic a year earlier: they increased their auto liability limits substantially and bought an umbrella policy. It wasn't because they were now "high risk." It was because they finally understood that the risk had always been there. They'd just been lucky.

Luck is not a plan.

Liability is about severity, not frequency

If you use the frequency/severity matrix, liability sits firmly in the Insurance Zone: low frequency, high severity. Most people won't face a catastrophic liability claim. But if you do, it can be financially disfiguring.

That's why the right framing isn't "How likely is it that I get sued?" The right framing is "If something awful happens and I'm found responsible, how bad could the damages be?"

In a car accident with serious injuries, damages can include:

- medical costs (which can be enormous)
- lost wages
- long-term care
- pain and suffering (depending on jurisdiction and case specifics)
- legal fees and defense costs (sometimes inside, sometimes outside policy limits)

Even if you have modest assets today, you have future income. That matters. Many households believe, "We don't have enough to be worth suing." But liability claims don't require you to be rich; they require you to be responsible for harm. And if you're a stable earner, you are financially relevant.

The foundation layer: high limits on auto and homeowners/renters

Before umbrellas, start with the base. Umbrella policies typically require you to carry certain minimum liability limits on your underlying auto and home/renters policies. In the real world, umbrellas don't replace those policies; they sit on top of them.

A practical way to think about it:

- Auto and homeowners/renters are the first layer that absorbs most normal claims.
- The umbrella is the second layer that catches the catastrophic overflow.

So if your base layer is thin, the umbrella either won't be available or it won't attach properly.

For auto insurance, the most common underinsurance problem is low bodily injury liability limits. People focus on collision and comprehensive (the car) and treat liability (the humans) as an afterthought. But the car is the cheaper part of the accident in many severe cases. The injuries are where the numbers live.

For homeowners or renters insurance, personal liability is the quiet line item that can matter most. Slip-and-fall injuries, dog bites, accidental damage you cause to others, and various personal liability claims can all land here depending on circumstances and policy terms.

None of this is a reason to be scared of your own driveway. It's a reason to stop thinking of insurance as a monthly bill and start thinking of it as *a limit you are choosing*.

What an umbrella policy actually does

An umbrella policy provides additional liability coverage above the limits of your underlying policies (auto and home/renters). It's designed for the big, rare scenarios: the kind that could punch through your auto policy limits and start eating your savings.

Umbrellas are often sold in $1 million increments. That number can feel cartoonishly large until you picture a severe accident involving multiple people, or a lifetime-care scenario, or a claim with substantial legal complexity. The job of the umbrella is not to match what you think is "reasonable." The job is to keep you from being personally exposed to a claim that doesn't care what feels reasonable.

Umbrella policies also typically require you to maintain certain underlying limits. That requirement is not the insurer being picky. It's the insurer making sure you're not using the umbrella as a substitute for basic responsibility. The umbrella is for overflow, not for everyday structure.

The hidden benefit: umbrellas can protect future you, not just current you

Many households treat liability insurance as asset protection: "We have to protect what we've saved." That's true, but incomplete.

Liability also threatens *future income*. If a judgment exceeds your coverage, the consequences can involve wage garnishment, forced liquidation of assets, or long-term payment arrangements. The exact mechanics depend on state laws and the nature of the judgment, but the theme is consistent: a big liability loss can reach forward in time.

That makes umbrellas especially relevant for households in their prime earning years. You may not be "wealthy" today, but you're building the kind of future that becomes worth defending. Buying a forcefield before you feel rich is one of those quietly smart moves that looks boring right up until it looks brilliant.

Why liability risk rises with normal life milestones

Umbrella insurance is sometimes framed as something you buy when you have a mansion. In practice, the need often grows with ordinary milestones:

- *Teen drivers.* More driving exposure, more risk.
- *A pool or trampoline.* Fun magnets can also be injury magnets.
- *A dog.* Most dogs are sweet. Liability claims don't care about your dog's personality; they care about what happened.
- *Hosting more people.* Kids' friends, parties, playdates.
- *More assets.* Savings accounts grow, retirement accounts grow, home equity grows.

None of these are arguments against a good life. They're arguments for updating your protection as life expands.

Deductibles vs. limits: don't confuse the knobs

Households often tinker with deductibles and think they've handled risk. Deductibles matter for *property* claims (your car repairs, your roof damage). Liability claims are different. Many liability coverages don't work like "pay the first $1,000." The key control is the *limit*.

A high deductible can be a great move when you have cash reserves and want lower premiums for property coverage. But a high deductible does not replace high liability limits. The forcefield is built with limits, and the umbrella is essentially a limit extender.

So the household system has two jobs:

- Cash (emergency fund/sinking funds) to handle deductibles and smaller losses.
- Insurance limits (auto/home + umbrella) to handle catastrophic liability.

If you only do one, you leave the other side exposed.

The shopping strategy: how to raise protection without getting fleeced

Insurance pricing can feel arbitrary because it's partly data-driven and partly market-driven. Premiums rise and fall based on claims trends, local factors, and underwriting appetite. In many markets, consumers have seen increases, including for umbrellas. That doesn't mean umbrellas aren't worth it. It means you should shop intelligently.

A practical approach:

- *Start with structure.* Decide the liability limits you want on auto and home/renters. Then decide the umbrella amount.
- *Quote as a bundle.* Umbrellas often price best when the underlying policies are with the same insurer. Not always, but often.
- *Ask about required underlying limits.* If you don't meet them, the umbrella quote is meaningless.
- *Revisit after major life changes.* New teen driver, new home, new assets, new side hustle, new pet—these are insurance events.

Also: be honest in applications. It's tempting to minimize risk factors to get a lower premium. That can backfire. The time to discover your umbrella doesn't cover a situation because of misstatements is not after an accident.

The side-hustle and landlord problem: when your normal policy stops being normal

Modern households increasingly have side income: gig work, consulting, Etsy shops, short-term rentals, a rental property, even occasionally driving for a delivery app. These can create coverage gaps. A personal auto policy may not cover certain business uses. A homeowners policy may have exclusions for business activity. A landlord situation changes liability exposure.

This doesn't mean you can't do those things. It means your insurance needs to match your life, not your self-image as "just a regular person." Regular people have complicated income now. The forcefield should be built around the truth.

Actionable takeaway: the 20-minute ``forcefield check"

Do this once this month:

1. Pull your auto and home/renters declarations pages (the summary pages that list coverages and limits).
2. Find your liability limits. Write them down.
3. Ask one question: *If we caused a severe injury accident tomorrow, would these limits protect our savings and our future income?*
4. If the answer is "I don't know" or "probably not," request quotes for higher underlying limits and a $1 million umbrella policy. Ask the agent or insurer what underlying limits are required.

Then pair it with a second move that makes the whole plan real: set aside the cash to cover your deductibles without debt. A forcefield works best when it sits on top of a household that can handle the smaller hits with cash and uses insurance only for the events that could change the rest of your life.

That's what "sleep well at night" coverage actually buys: not the fantasy that nothing bad happens, but the confidence that if something does, it won't take your future with it.

Chapter 5

Family, Future, and Resilience

Raising a family is the ultimate balancing act, where the joy of milestones often competes with the anxiety of bills. From the staggering 'peak costs' of childcare to the quiet pressure of supporting aging parents, the modern household faces challenges that love alone cannot solve. This chapter moves beyond basic budgeting to build a financial fortress around your family, ensuring that whether you are planning for college or preparing for the unexpected, your finances remain the steady ground beneath your feet.

5.1 The Economics of Childcare: Surviving the Peak Years

The day you realize childcare is not a "monthly expense" so much as a *temporary surcharge on your entire life* is usually the day you stop feeling mildly anxious and start feeling personally attacked by your checking account. You can be the kind of household that meal-plans, refinances, buys used cars, and still watch thousands of dollars vanish into the care-and-feeding of a child who cannot yet reliably locate their own shoes. It's not that you're bad with money. It's that ages 0–5 are a concentrated cost event—a financial weather system—and pretending it's "just another bill" makes it harder than it has to be.

The cruel part is how childcare collides with the exact moment many careers become most sensitive to continuity. Early childhood is often when promotions start stacking, when your reputation becomes an asset, when your income is finally climbing fast enough to make past sacrifices feel rational. Childcare charges you admission to keep that compounding machine running. So the real question isn't "Is daycare worth it?" The real question is: *What is the cost of buying time and career continuity during the peak years, and how do we keep the household from breaking while we do it?*

Childcare is also maddeningly local. Two friends can have similar incomes and similar kids and live in the same state, and one pays a figure that feels like a small mortgage while the other pays something that sounds merely unpleasant. The variation comes from the age of the child (infant care typically costs more than preschool), the setting (center-based versus home-based), and the county you're in. That's why childcare advice that starts with "the average cost is..." often ends with someone yelling at their phone. Your household doesn't pay an average; your household pays a price set by a small, capacity-constrained market in your zip code. The only useful conversation is the one that takes your actual number and runs it through real math.

Start with the number that matters most: your *net cost of childcare,* not the sticker price. Net cost is the check you write minus the tax help you actually capture and minus any employer benefits you actually use. If you don't calculate net cost, you'll talk yourself into a decision based on an inflated figure, then spend years paying for it in a different currency: stalled earnings, lost benefits, and the quiet erosion of financial resilience.

Here's what makes the net-cost calculation tricky: childcare creates *two* kinds of taxes in your life at the same time. On one side, there are the tax breaks designed to soften the blow (the Child and Dependent Care Credit and, for many people, dependent care flexible spending accounts through work). On the other side, there are the extra taxes triggered by having more earned income (payroll taxes, federal and state income taxes, possibly higher marginal brackets, and phaseouts of certain credits). When people say, "Basically my whole paycheck goes to daycare," they're often mixing three things together: the daycare bill, the taxes they pay to earn that paycheck, and the benefits they lose if they reduce income. Untangling those threads is the start of sane decision-making.

A practical way to begin is to stop asking "Is one salary eaten by daycare?" and instead ask a more honest question: *What does the second income contribute to the household after childcare and the costs required to earn that income?* Those work-required costs include commuting, parking, higher takeout spending during crunch weeks, work clothes, and the invisible stuff like the occasional paid convenience that keeps you functional (a cleaner twice a month, grocery delivery fees, a slightly more expensive apartment closer to work). This isn't moral failure. It's the real operating cost of a two-working-parent household.

Consider a concrete case. Maya and Jordan have a toddler and an infant. Maya earns $92,000; Jordan earns $68,000. Their center-based childcare quote for two kids is $3,200 a month

($38,400 a year). They live in a state with income tax. If Jordan considers stepping out of the workforce for a few years, they should not compare $68,000 to $38,400 and declare it "not worth it." That comparison is a cartoon.

A more useful back-of-the-envelope looks like this:

- Jordan's gross pay: $68,000
- Subtract estimated taxes on that income (federal, payroll, state; rough-all-in assumption for many middle-income households might land in the 20–30% range depending on the full household picture): say $16,000–$20,000
- Subtract childcare attributable to having both parents working: $38,400
- Subtract work-required costs (commuting, parking, higher convenience spending): say $3,000–$6,000

Even before any tax credits or benefits, you can see why the story feels brutal: the "keep working" surplus might be slim in the tightest years. But here's the twist that changes the decision: Jordan's paycheck isn't only buying cash flow. It's buying *future earning power*, continued retirement contributions, health insurance access in some households, and the ability to weather a job loss without instantly becoming a one-income family by necessity rather than choice.

The long-term math is where households either protect themselves or accidentally sandbag their future. Earnings compound. So do raises. If someone steps out for three to five years, the cost is often not "five years of salary." It can be *a permanently lower trajectory*—a smaller base salary when they return, slower promotion timing, fewer high-leverage projects, and a résumé that makes employers bargain harder. Even in fields where re-entry is common, the first job back is frequently a pay cut that takes years to climb out of. The decision may still be right for your family. But it should be made with eyes open: you are trading today's childcare bill for tomorrow's reduced compounding.

There's also retirement. Missing a few years of contributions is not just "less saved." It's lost employer match (free money), lost tax advantages, and lost time in the market. When money is tight, retirement is often the first thing to get "paused," because it doesn't scream. But retirement is the quiet bill that becomes an emergency later. If the childcare years force hard choices, an underrated win is to keep even a minimal retirement contribution alive, because it keeps the habit, captures any match, and preserves momentum.

Now, childcare does offer legitimate relief valves, and it's worth using every one you qualify for because these benefits are among the rare policies that directly subsidize the cost of being employed while parenting. Two common tools are: (1) a dependent care flexible spending account (DCFSA), if your employer offers it, and (2) the Child and Dependent Care Credit. Both are full of rules, caps, and coordination limits that make them feel like they were designed by someone who dislikes joy. The important thing is not to memorize every line of a tax form. The important thing is to treat childcare tax benefits like a discount you have to *activate*.

A DCFSA lets you set aside pre-tax dollars to pay for eligible dependent care expenses, up to an annual limit set by law and your plan rules. The key advantage is that you avoid paying certain taxes on that portion of income. For many households, that beats the credit. But you generally can't double-dip the same dollars for both the account and the credit; the benefits interact. This is why "we used the FSA and then got the credit" sometimes turns into "we used the FSA and then were surprised at tax time." The right move depends on your income, your tax bracket, and the size of your childcare bill relative to the limits.

Even if you don't want to turn your dinner table into a tax strategy summit, you can still do one simple thing that pays: before the year begins (or during benefits open enrollment), estimate your eligible childcare spending and decide whether a DCFSA is worth funding. If you're married and both of you work, the intent of these policies is literally to support that reality. Don't leave the support on the table because the paperwork is annoying.

But the bigger survival skill isn't tax optimization. It's *cash-flow engineering*. Childcare has an evil habit: it's due every month, in full, on schedule, and it does not care that your car battery died or your kid got strep again and you missed work. So you have to build a household system that treats childcare the way businesses treat payroll: as a non-negotiable operating expense that gets funded before anything that can be delayed.

One approach that works well in practice is to create a "peak-years budget" that is deliberately different from your "normal life" budget. This is a psychological trick with real power. If you try to maintain your pre-kid lifestyle *and* pay childcare, you'll feel like a failure every month. If you intentionally rename this season—"the peak years"—you can plan like a grown-up in a temporary crunch rather than a person who is perpetually behind.

A peak-years budget typically does three things:

First, it treats childcare like housing: paid first, automated if possible, protected from accidental spending. If the provider allows autopay, use it. If they offer a discount for paying monthly rather than weekly, take it. If there's a fee for credit cards and you're carrying a balance, don't turn your daycare bill into revolving debt. The goal is not points. The goal is not playing games with float. The goal is staying out of the debt spiral that childcare can trigger when one "small" shortfall turns into a carried balance with interest.

Second, it narrows the household's discretionary "blast radius." When money is tight, households often try to cut everything evenly and end up cutting nothing meaningfully. Instead, choose a few categories to aggressively simplify for a defined window: fewer trips, a smaller entertainment budget, slower home upgrades, a pause on expensive hobbies. Not forever. Just long enough to keep the peak years from borrowing against your future with credit cards.

Third, it actively buys back time in cheap ways. This sounds backwards—you're already paying for childcare, why spend more? Because time scarcity is the hidden accelerant that turns "tight but okay" into "chaos." Small, targeted spending can prevent expensive mistakes and burnout. A meal kit for one month during a work crunch might be cheaper than the DoorDash habit you develop when you're exhausted. A standing grocery order might prevent the $90 "emergency run" that somehow includes three unnecessary items and a toy. The point isn't to spend more. The point is to spend *strategically* so you don't spend *accidentally*.

The hardest part of childcare economics is that the decision is not purely financial. Some parents truly want to be home during these years, and no spreadsheet should bully them out of that. Some parents truly need to keep working for mental health, identity, or stability, and no cultural script should shame them into leaving. But even when values lead, math keeps you safe. The question becomes: *If we choose option A for emotional reasons, what financial scaffolding do we need so it doesn't become a crisis?*

If one parent stays home, the household needs to act like a one-income household on purpose, not by accident. That means redesigning fixed costs. Fixed costs are the bills that show up whether you had a good month or not: housing, cars, insurance, minimum debt payments, subscriptions. Fixed costs are what turn a job interruption into a catastrophe. Childcare years are a perfect time to conduct what you could call a "floor budget" exercise: identify the minimum monthly amount you need to keep the lights on, keep the household insured, keep food in the fridge, and keep debts from default. Then compare that to one income. If the floor budget can't fit inside one income with breathing room, you're not making a "stay home" choice; you're making a "we hope nothing goes wrong" choice.

If both parents keep working, the household still needs a version of that floor budget, because childcare increases the stakes of job loss. A layoff when you have childcare bills is different from a layoff when you don't. Your burn rate is higher, and your flexibility is lower because you may not be able to instantly drop care slots without losing them. In many markets, getting back into a good provider is not a simple "call them when you're ready." It's a waitlist, a deposit, and a stressful scramble. That reality changes the emergency fund conversation from "three to six months is nice" to "how many months do we need to keep childcare stable while we job-search without detonating our routines?"

That leads to one of the most counterintuitive childcare survival rules: sometimes you keep paying for childcare *even if one parent is temporarily not working*. If someone is between jobs for a month or two, keeping care can be the difference between an organized job search and a daily wrestling match with a toddler during interview calls. It can also preserve the slot. Financially, it may look irrational in the moment. Practically, it can be the move that protects the family's ability to get back to full income quickly. The decision depends on your savings and the availability of care in your area, but it's worth naming explicitly so you don't feel "wasteful" for doing what is strategically sensible.

Now for the part nobody wants to say out loud: childcare expenses create marital friction because they're emotionally asymmetrical. One person sees the bill and thinks, "We're paying strangers to raise our kid." The other sees the bill and thinks, "We're buying the ability to keep our life intact." Both reactions can be sincere. Both can be loving. And both can become toxic if the household treats childcare as an argument about values rather than a shared problem to solve.

A surprisingly effective tool is to reframe childcare as a *bridge* rather than a verdict. It's not "we're daycare people" or "we're not daycare people." It's "we are crossing ages 0–5 with the least long-term damage possible." When you frame it that way, you start looking for bridge-building tactics instead of ideological wins: shifting work hours to reduce paid hours, using part-time care plus grandparent coverage if that relationship is healthy, stacking summer camps and school-year programs strategically once public school begins, or switching providers when a child moves from infant to toddler rooms and the pricing changes.

And yes, the "public school raise" is real. It's not a raise in income; it's a drop in one of your biggest bills. But treating it like a raise changes behavior. Many households let the childcare bill disappear and then wonder why they still feel broke. If you plan in advance, the end of full-time childcare can fund priorities that were starved: rebuilding an emergency fund, catching up on retirement, paying down high-interest debt, or saving for the next lumpy cost (summer care, after-school programs, braces, or the inevitable car replacement you postponed).

Here's a practical way to capture the public school raise without requiring saint-level discipline: before childcare ends, decide what percentage of that "freed" money will be automatically redirected. For example, if you were paying $2,000 a month for one child and it drops to $600 for aftercare, you just "found" $1,400 a month. You could decide: $700 goes to retirement and emergency savings automatically, $400 goes to debt payoff, $300 goes back into lifestyle. The numbers will differ, but the principle is that you convert a disappearing bill into a deliberate plan rather than a spending leak.

One more case study, because the decision is rarely clean. Elena is a nurse working three 12-hour shifts; Sam is an office manager working a standard weekday schedule. They have one child, age two, and a second on the way. Full-time center-based care would cost enough that Elena's first instinct is to quit. But when they map out their actual schedules, they realize Elena's shift pattern can cover two weekdays of childcare needs without paying for it. They find a part-time home-based provider for the remaining days, and Sam negotiates one work-from-home day to reduce pickup chaos. Their solution isn't "cheap." It's *engineered*. They reduce paid childcare hours without erasing Elena's career, they keep benefits, and they preserve the option to adjust again when the second child arrives. The point isn't that everyone can do this. The point is that the best childcare plans often look less like a single decision and more like a custom schedule that gets revisited every six months.

That's the real survival mindset: treat childcare as a living system, not a one-time purchase. Your baby becomes a toddler. Your toddler becomes a preschooler. Prices change by age group. Your work situation changes. A grandparent's health changes. A provider closes. You don't need a perfect plan. You need a plan that can flex without collapsing the household.

A concrete takeaway to apply this week: take one hour, pull up your last month of bank and card transactions, and write three numbers on paper: (1) your actual monthly childcare cost, (2) the best estimate of your work-required monthly costs (commuting, parking, convenience spending you wouldn't have otherwise), and (3) the amount your household could cover if one income disappeared for 60 days. Those three numbers turn childcare from a vague cloud of stress into a solvable engineering problem—and once it's an engineering problem, you can build your way through the peak years without sacrificing the future.

5.2 The Sandwich Squeeze: Caregiving and Career Resilience

At 2:13 p.m. on an ordinary Tuesday, your phone rings. It's not a scam. It's your mom's neighbor. She fell on the front steps, the ambulance is on the way, and someone needs to meet her at the hospital. In movies, this is the scene where the adult child drops everything and rushes over, coat half on, heart pounding, love doing what love does.

In real life, you're also thinking: *Who's picking up our kid? What happens to my shift? Do I have enough PTO? If I miss another day, does my boss start quietly rewriting my future?*

That's the sandwich squeeze: you're pressed between people you'd take a bullet for, and the pressure comes from both sides. It isn't only emotional. It's logistical. And financially, it's a master class in why two-income households can feel stable right up until the moment they don't. When life demands caregiving hours, the family doesn't just lose time. It can lose income, health insurance, career momentum, and the ability to keep fixed costs from turning predatory.

Caregiving is also more common than many families like to admit out loud, because it sounds like something that happens to "older" people, or to "other" families. But plenty of households are in it while still raising kids, still paying student loans, still hoping the minivan can make it one more year. The squeeze often arrives in installments: first a few doctor visits, then a medication management routine, then a scary fall, then a diagnosis that rewires the calendar. By the time you call it caregiving, you've already been doing it for months.

The trap is that caregiving doesn't politely announce itself as a budget category. It arrives disguised as "just helping out," and then keeps escalating until you're rearranging your job around an unpaid second job you never applied for. The money leaks are sneaky at first: gas for extra driving, co-pays, a few takeout dinners because you're too tired to cook. Then they get loud: paying for home care a few hours a week, taking unpaid leave, cutting hours, turning down a promotion because you can't travel. This is how a financially careful household can end up in a high-stakes position without ever making a dramatic decision.

The antidote isn't panic. It's preparedness that feels almost boring in the moment and almost heroic later. The goal is career resilience: the ability to absorb a caregiving shock without your household becoming a one-income family by accident and without your long-term earning power quietly flatlining.

A deceptively powerful tool for this is the *Single-Income Stress Test*. It's not a spreadsheet to impress anyone. It's a question that forces the truth into the room:

> If one partner's income disappeared for a few months—or was cut in half—could the household still pay for the basics without burning the future?

Notice what's hiding inside that question. It's not only "could we survive?" It's *how*. Would you survive by pausing retirement contributions? By carrying credit card balances? By skipping insurance premiums? By draining your emergency fund and then living exposed? The test reveals not just whether you can make the numbers work, but whether your survival plan is sturdy or fragile.

Immediately after you run the test once, you can see the central problem of the sandwich squeeze: essential expenses are stubborn. Mortgages don't care that your dad's memory is slipping. Car payments don't adjust to your sister's chemo schedule. If caregiving forces a cut in income, the gap is where stress becomes structural.

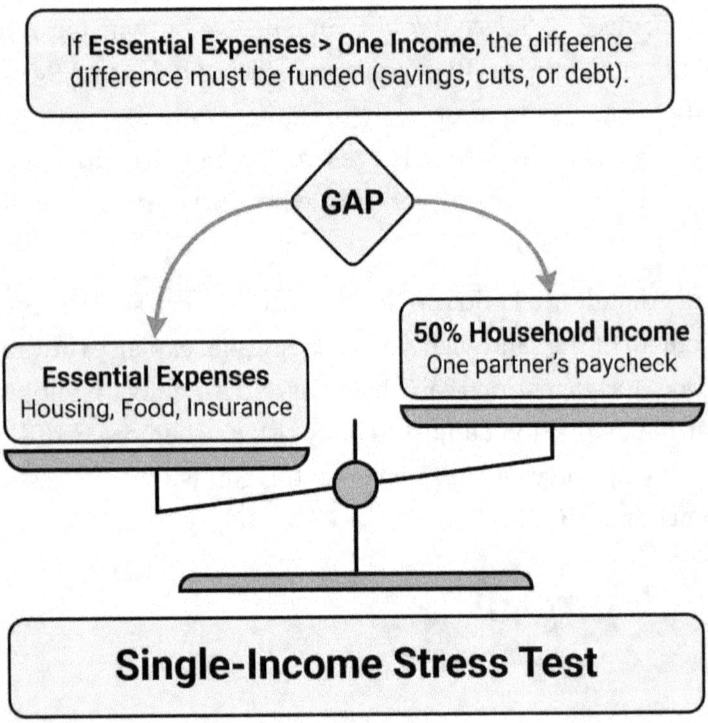

Once you see the gap, the rest becomes practical. If essential expenses are \$6,500 a month and one paycheck reliably nets \$4,500, then your caregiving plan is not "we'll figure it out." Your caregiving plan is a specific \$2,000 monthly gap that will be paid by something. The only question is *what* pays it: savings, reduced expenses, temporary income sources, or debt.

This is where many households accidentally choose debt, not because they're reckless, but because debt is frictionless and immediate. Caregiving has an urgent, morally persuasive flavor. Nobody wants to tell a parent, "I can't be there because the APR on our Visa is 24%." So the card comes out, and the household quietly starts borrowing against tomorrow to survive today.

The Single-Income Stress Test is how you prevent that quiet slide. It pushes you toward pre-decisions, and pre-decisions are what save you when emotions are high.

A realistic caregiving plan has three layers: cash-flow capacity, workplace strategy, and family logistics. The trick is that each layer supports the others. If you ignore one, the whole thing becomes wobbly.

Start with cash-flow capacity. This is not the same as an emergency fund. Emergency funds are for surprises. Caregiving is often a *semi-predictable* drain that escalates. You may need both: an emergency fund for shocks (the fall, the hospital stay) and a "caregiving runway" for the ongoing phase (home care, reduced work hours, travel, medication management). Even a small runway changes the emotional dynamics of the family, because it turns "helping" from a financial threat into a planned expense.

The household also needs to know which costs are truly essential and which are "essential because we're used to them." That's a tough distinction. The streaming services feel essential when you're exhausted. The second car feels essential until you price rideshare against a car payment. The premium phone plan feels essential until you realize it's a legacy decision from a time you were too busy to renegotiate. Caregiving seasons often force a household to re-rate its own spending in a more honest light. The goal is not austerity as a personality. The goal is to lower the household's fixed-cost floor so caregiving doesn't automatically create debt.

Now the workplace strategy. Caregiving can scramble a career not because you're less talented, but because employers reward reliability and predictability, and caregiving can erase both. Many households treat workplace policies like a dusty HR binder until the day they urgently need them. That's too late. Career resilience means knowing your options before the crisis hits.

If you have access to job-protected leave, flexible scheduling, remote work arrangements, or the ability to take intermittent leave in blocks of time, those tools can keep you employed while you care. Even when leave is unpaid, job protection matters because it preserves the relationship with your employer, your benefits, and your path back to normal hours. The tactical move is to gather the facts early: how eligibility works, what documentation is required, how health insurance premiums are handled during unpaid leave, and whether your role can realistically accommodate intermittent absences.

The emotional move is harder: you have to stop treating flexibility as a personal favor and start treating it as a business arrangement. You are not asking for kindness; you are proposing a plan that keeps you productive during a difficult season. Managers are human, but they also run teams. The caregiver who shows up with a clear proposal ("I can be fully available 9–2, I can take calls after 8 p.m., I can shift deadlines on these projects but not those") is easier to support than the caregiver who can only offer uncertainty. You may not get everything you ask for, but clarity gives you leverage.

Finally, family logistics. This is where caregiving gets political, even in loving families. Siblings have different capacities, different incomes, different histories. Some are helpful, some are avoidant, some are controlling, and most are just tired. If you don't assign roles deliberately, the most responsible person becomes the default caregiver, and that person's career often pays the price.

A simple but surprisingly effective practice is to turn caregiving into a *shared operating system*. Who is the medical point person? Who manages insurance paperwork? Who does weekly check-in calls? Who handles bills? Who can physically be there on weekdays? If those questions aren't answered, they'll be answered by inertia, and inertia is usually unfair.

Here's a concrete case study that shows how this plays out.

Tanya and Eric have two kids (ages six and nine). Tanya works in marketing with a manager who prizes responsiveness; Eric is a union electrician with steady hours and good benefits. Tanya's dad, Frank, lives forty minutes away. After a mild stroke, Frank can't drive for a while and needs help with medication, meals, and follow-up appointments. Tanya starts helping "a little." Within six weeks, she's missing morning meetings twice a week, her work is slipping into nights, and she's constantly on edge. Their spending rises: more takeout, more paid after-school coverage, more gas, and a few unexpected medical supplies. They feel busy, but not in control.

They run the Single-Income Stress Test and get a jolt. If Tanya's job went sideways, Eric's income would cover the mortgage, utilities, basic groceries, and insurance, but it would not cover everything else. Their margin of safety is much thinner than it feels. That changes the conversation: Tanya's caregiving can't be an improvisation anymore, because the household can't afford for her career to erode slowly.

They re-engineer the situation in three moves. First, Eric takes over all appointment driving that can be scheduled in advance, because his workday is less "meeting fragile" than Tanya's. Second, Tanya becomes the paperwork and medical portal point person, because she can do that work in defined blocks without vanishing during business hours. Third, they hire a home health aide for two afternoons a week for eight weeks—not forever, just long enough to stabilize Frank's recovery and give Tanya room to protect her job performance. The aide feels expensive, but the math is clear: it's cheaper than Tanya losing income, cheaper than credit card interest, and cheaper than the marital stress that comes from feeling trapped.

Notice what they did not do: they did not heroically absorb everything. They treated caregiving as a temporary project with a budget and roles. That's what resilience looks like.

The "caregiving runway" doesn't have to be large to change outcomes. Even $1,000 set aside specifically for care-related costs can prevent a high-APR spiral. A few thousand can buy paid help during peak intensity moments, which can keep a job intact. This is one of the rare times where spending money can *reduce* financial risk, because it preserves earning power.

It's also worth naming a hard truth: some caregiving costs are not just expenses; they are *exposure*. When you start paying for a parent's needs, it can be unclear where it ends. If a parent has limited savings, you can quietly become the long-term care plan. Plenty of adult children do this willingly, lovingly, and with pride. But financially, you need guardrails, or your household becomes the insurer of last resort.

Guardrails can be gentle. They can sound like: "We can cover $X per month, and we can do Y hours per week, but beyond that we need to look at other options." The point is not cruelty. The point is sustainability. A household with kids still at home cannot promise unlimited support without putting their own stability at risk, and instability helps no one.

If you feel a wave of guilt reading that, it's because guilt is the fuel of the sandwich squeeze. Guilt says: if you cared more, you'd do more. But money and time are finite. The responsible move is to decide, in calm moments, what you can do and what you cannot do, and to communicate that early enough that everyone can plan.

On the career side, one of the most strategic actions you can take is to identify which partner's job is more "shock absorbent." Some roles tolerate unpredictable absences better than others. Some have stronger legal or contractual protections. Some are more remote-friendly. Some are easier to reenter after a break. This isn't about valuing one person's career more as a human being. It's about recognizing that households run on cash flow, benefits, and optionality. If caregiving is likely, it can be wise to protect the more fragile or higher-growth career track, while the more stable role carries more of the scheduling burden. That choice can feel counterintuitive, but it can keep the household's long-term earning power healthier.

A related move is to build "career liquidity" before you need it. Career liquidity is the ability to pivot without panic: maintaining a current résumé, keeping certifications active, staying connected to professional networks, and knowing what jobs you could realistically get within 30–60 days if something changed. People do this when they fear layoffs; caregivers should do it because caregiving can force decisions just as fast.

There's also a small, unglamorous piece of resilience that matters enormously: consolidate information. Caregiving becomes a time sink not only because of the care itself, but because everything is scattered. Login credentials. Insurance cards. Medication lists. Doctors' names. Appointment histories. The adult child becomes a human search engine.

When you have a single document (digital and secure) that lists medications, providers, diagnoses, insurance details, and key contacts, you reduce the number of hours caregiving steals from workdays. That's not a productivity hack; it's a way to stop caregiving from consuming the calendar through administrative chaos.

The last piece is emotional, and it shows up financially: couples need to be on the same side. Caregiving can create a quiet competition of suffering. One partner carries more of the emotional labor; the other carries more income pressure. Resentment grows when each person feels unseen. Money becomes the proxy battlefield: "We can't afford that," versus "You just don't want to help." The fix is to name the shared objective: protect the parent, protect the kids, protect the household's stability. Then treat spending decisions and work decisions as part of that shared objective, not as evidence of love.

A concrete takeaway you can apply immediately: run the Single-Income Stress Test with real numbers this weekend. Write down

- one partner's monthly take-home pay,
- your essential monthly expenses, and
- the gap.

Then choose one action that reduces fragility within 14 days: refinance or renegotiate one fixed bill, set up a dedicated "caregiving runway" savings bucket with automatic transfers (even small), or schedule a meeting with HR to understand your leave and flexibility options before you're forced to ask under pressure. If caregiving knocks on your door, you'll still feel it in your chest—but it won't be able to knock your finances off a cliff.

5.3 Future-Proofing: Buckets for Education and Big Dreams

A certain kind of parental guilt has a calendar reminder built into it. Your kid is happily eating cereal, you're half-reading an email, and suddenly you remember that college is not a vague concept-it's a date. A year. An admissions season. A number with commas. And because you love this person more than your own comfort, your brain leaps straight from "they're in kindergarten" to "we're already behind."

That panic is understandable. It's also financially dangerous, because it can push good households into a backwards priority order: funding future tuition while underfunding their own ability to retire, withstand job shocks, or handle the next family emergency without debt. College is optional in a way your basic solvency is not. That's not a judgment about education; it's a statement about physics. You can borrow for school. You cannot borrow for retirement in the same way, and you certainly can't borrow at reasonable rates when you're older, tired, and out of runway.

The cleanest way to defuse the guilt without abandoning the dream is to use buckets-separate pools of money with separate jobs. When money has one big undifferentiated pile, every goal fights every other goal, and the loudest emotion wins. Buckets change the conversation. They turn "Should we save for college or retirement?" into "Which bucket gets filled first, and by how much?" That sounds like semantics until you try it. Then it feels like someone finally handed your household a steering wheel.

The most important bucket rule is the airline rule: put on your own oxygen mask before assisting others. In money terms, that means your household needs three things before it starts aggressively chasing tuition: (1) a workable cash buffer so normal life doesn't end up on a credit card, (2) protection against catastrophe (insurance and an emergency fund that matches your risk), and (3) retirement contributions that keep pace with the reality that nobody is coming to sponsor your older years. After those are in motion, education savings stops being a threat to your future and becomes what it should be: a generous, values-driven goal.

People sometimes push back: "But I don't want my child to have debt." Fair. But the worst-case scenario isn't "child has student loans." The worst-case scenario is "child has student loans and parents can't retire," which often becomes "child has student loans and also ends up supporting parents." That is the two-income trap with a time delay. Protecting retirement is not selfish; it's a gift that prevents future dependence.

Buckets also make room for a truth families rarely say plainly: the definition of "help with college" can be flexible. You might aim to cover two years at an in-state public university. You might aim to cover room and board but not tuition, or vice versa. You might aim to fund community college plus a transfer plan. You might aim to cover books and a laptop and travel home on holidays. Those are real forms of support. They can add up to tens of thousands of dollars and still leave your retirement intact.

The next step is choosing which bucket gets which tool. A bucket is a purpose; an account is the container. The mistake is to pick an account first because it has a shiny tax benefit and then force your goal to fit the account's rules. Instead, decide what you want the money to do, when you need it, and how flexible you want to be if life changes.

For education, the most famous tool is the 529 plan, sometimes called a qualified tuition program. The appeal is straightforward: you invest money, it can grow, and if you use it for qualified education expenses, the earnings can come out tax-free at the federal level (and often with state tax benefits depending on where you live and how your state's rules work). But "qualified education expenses" is one of those phrases that sounds obvious until you discover it's not universal. Different education tax benefits use similar language with different meanings, and mixing them up can lead to unpleasant surprises.

A useful mindset is to treat education tax benefits like overlapping but distinct maps. The IRS has multiple programs-credits, deductions, exclusions, and savings vehicles-and each has its own definition of which expenses count and under what conditions. If you ever find yourself stacking benefits (for example, using a 529 distribution while also claiming an education credit), you have to be careful not to pay the same expense with two tax-favored dollars. The government is happy to help with education; it is less happy to help twice for the same receipt.

The bucket approach helps here too. A 529 bucket can be your "college tuition and fees bucket." A separate savings bucket can be your "flexible education support bucket"-money you might use for a car, a computer, tutoring, a test-prep course, a move, an internship in a high-cost city, or the random costs that show up in young adulthood that aren't technically tuition but are very much part of launching a life. Flexibility is not inefficiency. Flexibility is resilience.

There's another tool that feels old-fashioned but can be surprisingly elegant for certain households: U.S. Savings Bonds, specifically Series EE and Series I bonds, which under certain conditions can have their interest excluded from income when used for qualified higher education expenses. It's not a universal fit—there are income limits and rule details that matter—but conceptually it offers a different flavor of education bucket: one that may feel safer to a household that's wary of market volatility right as college approaches. If you're the kind of person who lies awake worrying about a market drop during junior year of high school, you're not irrational. You're recognizing that time horizon changes the rules. The right tool depends on whether your primary fear is missing growth or missing certainty.

That time-horizon point matters no matter what account you use. Money for college is not one goal; it's a ladder of goals with different deadlines. Tuition for freshman year is due first. Tuition for senior year is due later. The money needed in the next three years should not be invested the same way as money needed in 12 years. Households get into trouble when they treat college savings like a single long-term investment, right up until the moment it becomes a short-term bill.

A simple way to make this real is to split the education bucket into "near" and "far." Near-term education money (say, within three years of when you'll spend it) belongs in low-volatility places: cash, high-quality short-term instruments, or whatever conservative option your 529 plan offers. Far-term education money can take more risk because it has more time to recover. This isn't about predicting markets. It's about acknowledging that markets are moody and calendars are not.

Now add a layer that makes education planning feel unfair: financial aid formulas. Two families can save the same amount and get different aid outcomes depending on how and where the money is held and whose name is on the account. Even if you don't expect need-based aid, it's worth understanding that "where the money sits" can matter. Recent FAFSA changes also shifted how some resources and distributions are treated. The headline version: account ownership and the flow of money can affect aid calculations, and the rules can change over time. The bucket approach doesn't solve policy complexity, but it makes it easier to adapt because your money is already organized by purpose.

This is also why a household should resist going all-in on a single education strategy when the future is uncertain. Your kid might get scholarships. They might choose a cheaper school. They might choose a more expensive one. They might not go to college immediately. They might go into a program where costs look different than you expected. A single rigid plan can become a stressor. Multiple buckets give you optionality.

Let's put this into a real-world case.

Marcus and Alina have two kids, ages four and seven. Their household income is solid, but not "throw money at problems" solid. They're still recovering from the childcare years: they kept working, but it came with higher convenience spending and a slower debt payoff than they'd planned. Marcus is now pushing hard to max out 529 contributions because his parents couldn't help him, and he doesn't want that story repeating. Alina is worried about their own retirement because she watched her aunt "borrow" from her kids emotionally and financially in her 70s.

They decide to build buckets with rules that are easy to follow on tired weekdays:

- *Stability bucket:* One month of expenses in a high-yield savings account, then build toward a larger emergency fund target. This bucket is not negotiable because it prevents credit card debt during normal chaos.

- *Retirement bucket:* Contribute enough to capture all employer match immediately, then increase contributions by 1% each time one of them gets a raise until they hit a comfortable percentage. They automate it so it happens without monthly debate.
- *Education bucket (529):* A set monthly amount per kid that feels modest but consistent. They choose an age-based investment option so the risk level automatically becomes more conservative as college approaches.
- *Dream bucket:* A separate savings account labeled "Big Dreams," which covers family travel and experiences and, importantly, acts as a pressure release valve so they don't resent the other buckets.

The emotional breakthrough is that they stop treating college as a single do-or-die goal and start treating it as one of several commitments. Marcus doesn't feel like he's abandoning his kids; he's building consistent support. Alina doesn't feel like she's sacrificing her own future; she's protecting it. The buckets turn what used to be a philosophical argument into a weekly system.

That system also prevents a common trap: using college savings as a proxy for love. When money is messy, people look for a "pure" goal that feels virtuous. College becomes that goal. But if you fund college by ignoring retirement, you're not building a virtue; you're building a future emergency with your signature on it. Buckets keep virtue from becoming self-sabotage.

There's another subtle advantage to buckets: they help you handle the "helping others" requests that show up when your own parents age. A household in the sandwich years may be asked to help with elder care while also saving for kids. Without buckets, every request feels like it competes with your child. With buckets, you can respond with something clearer: "We have $X per month available in our family support bucket," or "We can cover these specific expenses." Clear boundaries are kinder than vague promises.

Now, some households will say: "We can't fund all these buckets." That's not failure; it's information. If your budget can't support even minimal retirement plus minimal emergency savings plus minimal education contributions, the problem isn't your bucket system. The problem is that your fixed costs are too high for your income, or your income is too low for your obligations, or both. Buckets don't magic money into existence. What they do is reveal tradeoffs early enough that you can make changes with less pain: housing choices, car choices, debt payoff strategies, job changes, side income, or simply resetting expectations about what "help with college" can realistically mean.

When money is tight, the most powerful education move might not be a 529 contribution at all. It might be building the household's stability and credit health so you can say yes to opportunities later: a relocation for a better job, a training program that increases income, a move to a better school district, or the ability to cash-flow a portion of college without borrowing at high rates. Financial stability is the quiet sponsor of future options.

If you want one practical method to implement buckets without creating a bureaucratic mess, use a "two-account plus subbuckets" setup:

- One checking account for bills and daily spending.
- One high-yield savings account that allows named subaccounts (or separate savings accounts) for: emergency fund, near-term goals, and education support.

Then automate transfers on payday: retirement contributions first (via payroll), emergency fund second (until it hits your target), and education third. If you can't automate it, it won't happen consistently, and inconsistency is what turns "we're saving" into "we're stressed."

A final point that's unpopular but liberating: your kid's future is not solely financed by your savings. It's financed by your parenting, your stability, the example you set, and the economic flexibility you preserve. A household that models calm, consistent money habits gives a child an inheritance that doesn't show up on a statement: the ability to navigate adulthood without fear. That doesn't pay tuition directly, but it reduces the odds that your child repeats the same financial scrambles.

A concrete takeaway to apply this week: open or rename your savings buckets so the purpose is visible, then set one automatic transfer for each of these priorities in this order: (1) retirement contributions at least up to the employer match, (2) a small weekly transfer to an emergency fund until you reach one month of expenses, and (3) a monthly education contribution you can sustain for two years without touching credit cards. The amounts can be modest; the consistency is the win.

When the buckets are real and the transfers are automatic, the guilt gets quieter, and the dreams get more durable.

5.4 The ``In Case of Emergency'' Box: Essential Documents

There's a particular kind of panic that doesn't feel like panic until later, when you replay it and realize your hands were shaking the whole time. It's not the dramatic "we're in danger" panic. It's the administrative kind: you're in a hospital waiting room, or on the phone with an insurance company, or standing in your kitchen staring at a water-stained ceiling, and someone asks a simple question—*What's the policy number? Who's the beneficiary? Do you have the paperwork?*—and you realize your household has a lot of money systems and almost no money *instructions*.

Most families think they're unprepared for disasters because they don't have enough bottled water. The quieter truth is that families fall apart financially because they can't prove who they are, what they own, what they owe, and who is allowed to make decisions. The "in case of emergency" box is less about catastrophe and more about continuity. It's the difference between a bad week and a months-long unraveling.

If you're a parent, the stakes go up. It's not just "how do we pay bills if something happens?" It's: *Who takes the kids? Who can access money to feed them? Who has legal authority to make medical decisions? Does the right person even know where our accounts are?* These are uncomfortable questions, which is why they're often postponed until the universe forces them onto your calendar.

The good news is that you don't need a law degree or a vault in a mountain. You need a simple system built around the Big Three: *Wills, Beneficiaries, and Guardianship*. Everything else is supporting cast.

A will is the set of written instructions for what happens to your assets after you die. It can also name a guardian for minor children. People sometimes treat wills like a "rich people" thing, but the reason parents need them isn't yachts. It's kids. If you die without a will, state law decides who inherits, and the court process (probate) can be slower and more expensive than families expect—right when your household needs speed and clarity.

Beneficiary designations are the "who gets it" instructions built into specific accounts—like a life insurance policy, a 401(k), an IRA, and many brokerage accounts. Here's the part that surprises people: beneficiary designations often *override* your will. If your will says "everything goes to my spouse," but your retirement account still lists your ex from 12 years ago, the account may pay the ex. It's not a rare edge case; it's a common real-life mess. This is why the emergency box is not just "documents to store." It's "documents to periodically update."

Guardianship is the core parenting question: who raises your children if you can't? Many loving families assume "of course my sister would take them," or "my parents would step in." But assuming is not the same as naming. If you don't formally nominate a guardian, the court will decide. Courts do their best, but they are not mind-readers, and they are not your family historian. Your job is to make your intention legible.

Notice what all three have in common: they're not about money in the abstract. They're about *authority*. Who can act? Who can sign? Who can access? Who can decide? Financial resilience isn't only dollars; it's decision rights.

A real-world example makes this painfully clear.

Chris and Leila are both in their thirties with two kids under eight. They have normal life complexity: two jobs, a mortgage, daycare bills that recently turned into after-school care, a retirement plan through work, and a small life insurance policy they bought when their first child was born. They did a will online after the second baby—felt responsible, checked the box, moved on.

Then Leila is hospitalized unexpectedly. She recovers, but for two weeks she can't handle any paperwork, and Chris is suddenly doing everything while terrified. He tries to file a short-term disability claim and realizes he doesn't know which insurer handles it. He calls HR and gets a portal link. The portal asks for policy info and doctor certifications. He needs Leila's login credentials. He doesn't have them. Meanwhile, Leila's sister offers to help with the kids, but the school office wants written permission on file for pickups. Chris has to find the form and submit it while bouncing between hospital visits, work calls, and bedtime.

Nothing "financial" happened—no fire, no fraud, no death. But their lack of a document system turned an already hard moment into chaos. Afterward, they build a box. Not because they like paperwork, but because they never want to be trapped like that again.

That's the point: the emergency box is not for hypothetical disasters. It's for the kind of disruption that happens in real households: illness, an accident, a sudden work interruption, a move, an insurance claim, even just the brain-melt of grief. When your cognitive bandwidth is low, systems either carry you or crush you.

So what goes in the box?

Start with identity and access—the boring stuff that becomes priceless on the worst day.

You want a secure place (physical and digital) for copies of: birth certificates, Social Security cards (or at least the numbers recorded securely), passports if you have them, marriage certificate, and any legal name-change documents. If your home burns down or floods, proving identity is step one to everything else: insurance claims, replacing documents, getting access to accounts, securing housing.

Next is the money map: a one-page "if you need to run the household without me" sheet. This is not where you paste passwords in plain text. It's where you list *what exists* and *how to find it*.

Include: bank names, last four digits of key accounts, where paychecks are deposited, where bills are paid from, credit card issuers, mortgage servicer, auto lender, student loan servicer, and any other debt that would surprise a spouse if it went unpaid for 60 days. Add your employer info, payroll contacts, and where benefits live (HR portal, benefits administrator). If you have a safe deposit box, record which bank and where the key is. If you have a password manager, record how to access it and who the emergency contact is in that system. The goal is to make your household legible to someone who is panicked and sleep-deprived.

Then comes insurance, which is where families lose the most money through delays. Put in: health insurance cards and plan details, life insurance policies, homeowner's or renter's insurance declarations page, auto insurance details, and disability insurance (through work or private). Insurance is the financial shock absorber for the exact events that produce emergencies. But it only absorbs shock if you can file a claim efficiently.

Add property and proof. If you've ever watched someone try to remember what they owned after a theft or a disaster, you understand why a household inventory matters. Take a slow video walk-through of your home: open closets, pan drawers, show serial numbers on electronics, capture the "boring expensive" stuff (appliances, tools, furniture). Back it up somewhere not in your house. In a crisis, memory becomes unreliable; proof becomes power.

Now the legal heart of the box: wills, guardianship nominations, and powers of attorney.

A will handles what happens after death. But what about the more common scenario: you're alive and can't act? That's where *powers of attorney* come in. A financial power of attorney authorizes someone to handle financial matters on your behalf (paying bills, managing accounts) if you're incapacitated. A medical power of attorney (and related health care directives, depending on your state) authorizes someone to make medical decisions if you can't. Without these, spouses may be surprised by how hard it can be to get things done quickly, especially when dealing with institutions that are trained to say no unless authority is crystal clear.

Parents should also include a temporary guardianship letter or emergency child-care authorizations, if appropriate in your state and situation, so a trusted person can pick kids up from school or take them to the doctor if you're unreachable. Schools, doctors, and daycares are rightly cautious. A little paperwork now can prevent a nightmare later.

Finally, beneficiaries. This is the part that feels like it should be handled by the will, which is why it's so often wrong. Beneficiary forms are scattered across institutions, each with its own portal and its own version of "update." The emergency box should include a beneficiary checklist: every retirement account, every life insurance policy, every HSA, and any payable-on-death or transfer-on-death designations on bank and brokerage accounts.

You don't need to store every beneficiary form in the box if you can't easily obtain them, but you do need a record of where each account is held and when you last confirmed the beneficiary. If you can answer, "We checked beneficiaries in October 2025," you're leagues ahead of most households.

A subtle but critical parenting detail: if you name minor children as beneficiaries directly on some accounts, it can create complications, because minors generally can't legally control inherited assets without a custodian or court involvement. Many families use a trust or other structure, depending on state rules and their goals, to avoid a situation where a child receives a large sum at 18 with no guardrails. This is where an estate attorney can be worth real money: not for fancy wealth strategies, but for preventing accidental chaos.

The storage question matters too. The emergency box should be redundant because emergencies are rude. Think in two layers:

- *Physical:* a portable, fire-resistant box at home for copies and a few originals you truly need. Some originals may be better in a safe deposit box, but remember that safe deposit access can be tricky if the wrong person is authorized or if banks have limited hours during a crisis.
- *Digital:* encrypted, backed up, and accessible. A secure cloud folder with strong authentication, or an encrypted drive stored offsite, can keep documents available even if you can't get into your house.

The point is not to create a fortress. It's to avoid a single point of failure. If a storm knocks out power, you still want access. If you're traveling, you still want access. If your spouse is the one dealing with the crisis, you want them to have access without a scavenger hunt.

This is also where the emergency box intersects with everyday household systems. If only one partner knows where everything is, your household is fragile. The emergency box is the antidote to the "CFO spouse" problem, where one person carries the entire operational knowledge of the family. Love is not the same as redundancy. Redundancy is what keeps love from becoming a disaster plan.

To keep the box from becoming a one-time project that rots in a closet, tie it to a recurring trigger. The best triggers are life events you already have: taxes, open enrollment, a birthday month, daylight saving time, the month you renew car registration. Once a year, spend 30 minutes checking: have we moved? changed jobs? opened new accounts? had a new baby? lost a family member? changed who we trust? Any "yes" means the box needs updates.

One more reality check: the emergency box is not only for death and disaster. It's for the mundane emergencies that cost money because they cost time. The faster you can produce a document, prove a claim, or authorize a decision, the less likely you are to miss work, incur late fees, or accept bad terms because you're desperate. Preparedness is a financial strategy because speed is a financial advantage.

A concrete takeaway you can do in one weekend: create a one-page Household Money Map and store it in two places (one physical, one digital). The Money Map should list every account category (banking, credit cards, loans, insurance, retirement), the institution name, and the "how to access" method (website/app name, customer service number, where the password manager lives). Then schedule one 45-minute appointment—on the calendar, like it matters—to review and update beneficiaries on every retirement and life insurance account you can find. Those two actions alone turn your household from "we hope" to "we planned," which is the difference between an emergency and a financial freefall.

Chapter 6

The Investor's Toolkit

Investing often feels like a secret language designed to keep you out, but stripping away the jargon reveals a surprisingly simple toolkit. This chapter hands you the wrench and screwdriver-showing you how to build a portfolio that works for your family without needing a finance degree. We will move past the fear of the stock market and learn how to cook a balanced financial meal using just a few high-quality ingredients.

6.1 Risk, Return, and Volatility

The first time most households feel "investment risk," it isn't on a spreadsheet. It's on a random Tuesday at 9:47 a.m., between meetings, when someone checks an account balance on their phone and sees a number that is *lower than it was last week*. The stomach drops. The brain starts narrating: *What if this keeps falling? What if we're the kind of people who lose money? What if we did something irresponsible?* The market hasn't knocked on your door and demanded a payment, but your nervous system acts like it has.

That moment—more than any textbook definition—is why it helps to redraw the map. In real life, the biggest day-to-day "risk" of investing is often not the permanent loss of money. It's the emotional toll of watching prices swing, and the downstream damage those feelings can cause: panic-selling, second-guessing, abandoning the plan, or freezing in cash for years while inflation quietly eats your future. Volatility is the part that grabs you by the collar. It's loud, dramatic, and very good at making sensible people do expensive things.

Start by separating two ideas that get mashed together in everyday conversation:

Volatility is the rollercoaster—prices moving up and down, sometimes sharply, sometimes for no obvious reason. You can have volatility without permanent loss, the way you can have turbulence on a flight that still lands safely.

Permanent loss is the crash—money that doesn't come back because you sold at the wrong time, or because you owned something fragile that truly broke (a single company that went bankrupt, a concentrated bet that never recovered, a scam, a portfolio built on leverage, or a plan that forces withdrawals at the worst moment).

Most households don't ruin their future because the market had a bad year. They ruin it because volatility provoked a decision that turned a temporary drop into a permanent setback.

Here's the uncomfortable truth that makes the rest of investing make sense: the market doesn't pay you extra return for being smart. It pays you extra return for bearing discomfort. That discomfort is volatility—the feeling that the ground is moving under your feet while you're trying to build a life on it. If you want inflation-beating growth over decades, some amount of price swing is not a bug in the system. It's the price of admission.

Volatility is a fee, not a fine. A fine is what you pay because you did something wrong. A fee is what you pay because you want access. You don't pay a gym membership because you're being punished; you pay it because you want the benefits that come with showing up. Market volatility is the membership fee for participating in long-term growth. It's unpleasant, and it's not optional.

But a fee can still be too high for your budget. That's where the personal side of risk comes in, and it's much more practical than any chart of historical returns.

There are two kinds of "how much risk can we take?" and they're not the same question.

Risk capacity is your household's financial ability to endure a downturn without your life breaking. It's about math and timing: stable income versus shaky income, fixed expenses versus flexible expenses, emergency savings, insurance coverage, debt payments, how soon you need the money, and whether a job loss would collide with an investment decline.

Risk tolerance is your emotional ability to live through a downturn without bailing out. It's about psychology and biography: how you react to uncertainty, how you grew up around money, whether you've experienced layoffs, whether you equate account balance changes with personal failure, and whether you and your partner get calmer together—or amplify each other's anxiety.

A household with high incomes and a big cushion can have low tolerance. A household with modest income but iron nerves can have high tolerance. The best plan respects *both*. Ignore risk capacity, and you can be forced to sell at the worst time. Ignore risk tolerance, and you can choose to sell at the worst time. Either way, you're selling at the worst time.

A concrete example makes the distinction real.

Consider Maya and Jordan, a two-income household with a toddler and a mortgage. Maya works in healthcare; Jordan works in a tech role tied to the business cycle. They've been diligent: retirement accounts are invested mostly in stock funds because "stocks are best long-term," and they've heard that time in the market matters. Their emergency fund is thin because daycare and housing consume most of the monthly cash flow. Then a downturn hits. Tech companies start cutting. Jordan's role feels less secure right when their investments are down.

Notice what happened: their portfolio didn't become "bad." Their *timeline* and *cash needs* collided with volatility. The real risk isn't that markets fell. The real risk is that they might have to liquidate investments—locking in losses—because the household doesn't have enough cash resilience to ride out simultaneous income stress and market stress.

In that situation, the correct lesson isn't "stocks are dangerous." The lesson is "stocks are volatile, and volatility becomes dangerous when your plan requires cash at the wrong moment." The fix is often less dramatic than people assume: build a stronger cash buffer, reduce fixed monthly obligations where possible, align the investment mix with the time horizon, and stop treating retirement money like an emergency fund with a nicer interface.

Volatility is also where partnership dynamics show up. In a two-income home, one person often becomes the "market watcher" by accident—the one whose phone has the app, whose feed serves the scary headlines, who feels responsible for keeping the family from making a mistake. That responsibility can morph into control, or into silence, or into a cycle of frantic updates: *"It's down again." "How much?" "A lot."* The numbers become emotional weather, drifting into dinner conversations and bedtime routines.

A useful household rule: if a number can change by thousands of dollars without you doing anything, checking it daily is a form of self-harm. Not because information is bad, but because your brain is not a neutral observer. It's a storyteller, and it specializes in worst-case narratives.

The market's day-to-day motion is mostly noise. Yet it feels meaningful because it triggers ancient wiring: loss aversion (losses hurt more than gains feel good), recency bias (what just happened feels like what will keep happening), and action bias (doing something feels safer than doing nothing). Volatility doesn't just move prices; it moves people.

Now, about "return," because this is where households either get discouraged or get reckless.

Return is what you get paid for taking risk *that is actually compensated*. Over long periods, broad stock markets have historically delivered higher returns than cash-like holdings precisely because stock ownership is not guaranteed and not smooth. If returns were steady and guaranteed, they wouldn't be high; the world would arbitrage that away. The extra return is the risk premium—the payment for tolerating uncertainty and the possibility of being temporarily underwater.

But not all uncertainty is compensated. Betting your retirement on one company is uncertainty, sure—but it's often *unpaid* uncertainty. There's no special reward for taking a risk that could have been diversified away cheaply. Likewise, taking on high fees, leverage, or opaque products adds risk without necessarily adding expected return. Those are the kinds of risks that tend to turn volatility into permanent loss.

So when people say, "I'm not willing to take risk," what they often mean is, "I'm not willing to feel scared." That's honest. It's also solvable—not by eliminating volatility (you can't), but by designing a plan that makes volatility survivable.

A powerful way to do that is to match the *job* of the money to the *behavior* of the investment.

Money you may need soon—job transition funds, deductible buffers, near-term home repairs, tuition bills—should not be assigned to a vehicle that can be down 30% right when you need it. Volatility turns into an emergency when the clock is short.

Money you won't need for a long time—retirement decades away—can afford volatility *if* the household won't raid it when fear spikes. Here, the real challenge isn't market math; it's household governance: agreements, automation, and guardrails that protect the plan from your future stressed-out self.

That brings us to a definition of "risk" that's more useful at the kitchen table:

Risk is the chance you'll be forced—or tempted—to make a permanently damaging decision.

Volatility increases that chance by raising the emotional temperature. A plan reduces that chance by lowering the need for urgent action.

To make this less abstract, picture two families with identical portfolios and identical market returns. Family A has a tight emergency fund and high fixed costs. Family B has a robust emergency fund and lower fixed costs. When a recession hits and the portfolio drops, Family A panics because every dollar feels spoken for; Family B is annoyed but not threatened. Same market. Same portfolio. Different risk, because the household structure changes the likelihood of selling low.

This is why "risk" isn't just an investment setting. It's a household systems problem.

And yes, volatility can be brutal even when nothing is fundamentally wrong.

Imagine contributing steadily to a retirement account for years, only to watch the balance fall below where it was a year ago. Your brain does the simplest possible math: *I added money and I have less. That means I'm losing.* In reality, you may have bought shares at lower prices—setting up for future gains when prices recover. But emotionally, it feels like working overtime to dig a hole.

This is the paradox: the very feature that makes long-term investing work—being able to buy more shares when prices are down—feels like failure in the moment. It's not intuitive, and it doesn't feel good. That's why households need scripts and rules in advance, not motivational posters during the storm.

Here are a few scripts that actually hold up when numbers are red:

- *"A lower price is not a broken investment; it's a lower price."* (Broken is when the underlying asset is fundamentally impaired or you're forced to sell.)
- *"We don't change long-term plans based on short-term headlines."* (Headlines exist to activate emotion, not to improve your returns.)
- *"If we feel an urge to sell, we wait 72 hours and re-check our reasons, not our balance."* (A delay doesn't eliminate risk; it reduces impulsive mistakes.)

Those rules sound simple. In practice, they're powerful because they separate *feeling* from *action*. You're allowed to feel nervous. You're not allowed to convert nervousness into a permanent decision without a cooling-off period.

Another practical tool: stop describing volatility in percentages when talking as a couple. Percentages can feel abstract—until you translate them into the language of your life.

A 25% decline on $200,000 is $50,000. That's not "twenty-five percent." That's "a car," or "a year of daycare," or "six months of mortgage payments." The point isn't to scare yourselves. The point is to plan honestly. If a drop of that magnitude would cause you to abandon the plan, the issue isn't bravery—it's fit. Adjust the investment mix, increase the cash buffer, or reduce fixed costs so you're not depending on emotional heroics.

That's also where the word *time* becomes your best friend.

Time does two things. First, it gives investments room to recover from downturns. Second, it gives your household room to not care about the next twelve months. When your near-term needs are covered by cash and stable income, volatility becomes background noise instead of a household emergency.

Now, it's tempting to look for a "safe" alternative that still delivers high returns. That's a natural wish. It's also where a lot of financial pain begins—because salespeople and internet gurus are happy to offer you exactly that story.

Be suspicious of anything that promises equity-like returns without equity-like volatility. Sometimes it's a misunderstanding. Sometimes it's selective marketing. Sometimes it's hidden risk: illiquidity (you can't easily get your money out), credit risk (someone can't pay you back), leverage (borrowing magnifies losses), or complexity risk (you don't really know what you own). Many households get hurt not by boring, transparent market volatility, but by reaching for "smoother" returns that turn out to be smooth until they aren't.

A useful household filter: if you don't understand how something could lose money, you're not ready to own it.

Volatility, in contrast, is honest. You can see it. It doesn't pretend to be something else. It's the market saying, every day, "I might be up; I might be down." Your job is to build a plan that doesn't require you to guess which one it will be next week.

One more real-world pattern worth naming: people often say they want "less risky" investments when what they really need is "more cash." These are different. A low-volatility bond fund can still go down. A stable-value option can have restrictions. A bank savings account won't make you rich, but it's designed to be there on the day you need it. Cash is not an investment engine; it's a shock absorber for your life.

That's why emergency funds are not boring "personal finance advice" to check off and ignore. They are a volatility-management tool. They keep you from turning market swings into forced selling. They buy you time, and time is what makes volatility survivable.

If you want a single sentence that captures the whole game, it's this: *You don't have to predict markets; you have to prevent volatility from cornering you.*

So what does that look like in a two-income household that's trying to escape the trap?

It looks like building your plan around *decision points*. The market will hand you dramatic days. The question is whether those days force a decision.

- If one income disappears, do you have a cash runway long enough to job-search without raiding investments?
- If both jobs feel shaky in the same recession, do you have a "household recession plan" (expense cuts you can make quickly without chaos)?
- If the market drops 30%, do you already know what you will do? (Usually: keep contributing, rebalance if that's your rule, and avoid changes driven by fear.)

These aren't academic questions. They're the difference between volatility being a fee versus volatility being a fine.

And there's a subtle benefit to handling volatility well: it changes how money feels inside a relationship. When a household has a plan that can absorb shock, market drops become irritating rather than existential. The conversations change from "Are we in trouble?" to "Is our system working?" That shift is how families stay invested long enough for compounding to matter.

A concrete, actionable takeaway you can do this week:

Write a one-page "Volatility Plan" and agree to it together. Include (1) the maximum portfolio drop you're willing to tolerate without changing course (write it in dollars, not just percent), (2) the minimum cash buffer you will maintain to avoid selling investments in a downturn, and (3) a cooling-off rule for any investment change (for example: no allocation changes within 72 hours of a scary market day; discuss once, decide once, document the reason). Then set your account apps to hide daily percent changes if they trigger anxiety, and check balances on a schedule you can emotionally afford. This isn't about being fearless. It's about being un-panickable.

6.2 Understanding Asset Classes

A lot of investing advice accidentally sounds like this: "Pick a fund." As if your financial life is a single pot of money with one job.

Real households don't work that way. Your money has multiple jobs, often at the same time: keep the lights on if a paycheck disappears, show up intact for a down payment, grow into a retirement that won't be bullied by inflation, and do all of that without turning your kitchen table into a nightly crisis hotline. The fastest way to make investing feel sane is to stop thinking in terms of "good investments" and start thinking in terms of *ingredients*. Different ingredients behave differently under heat. You don't judge salt by whether it bakes like flour.

The three big ingredients—stocks, bonds, and cash—are the core of most household portfolios because they each do something distinct. When you combine them well, you get a plan that can feed you in good markets *and* bad ones.

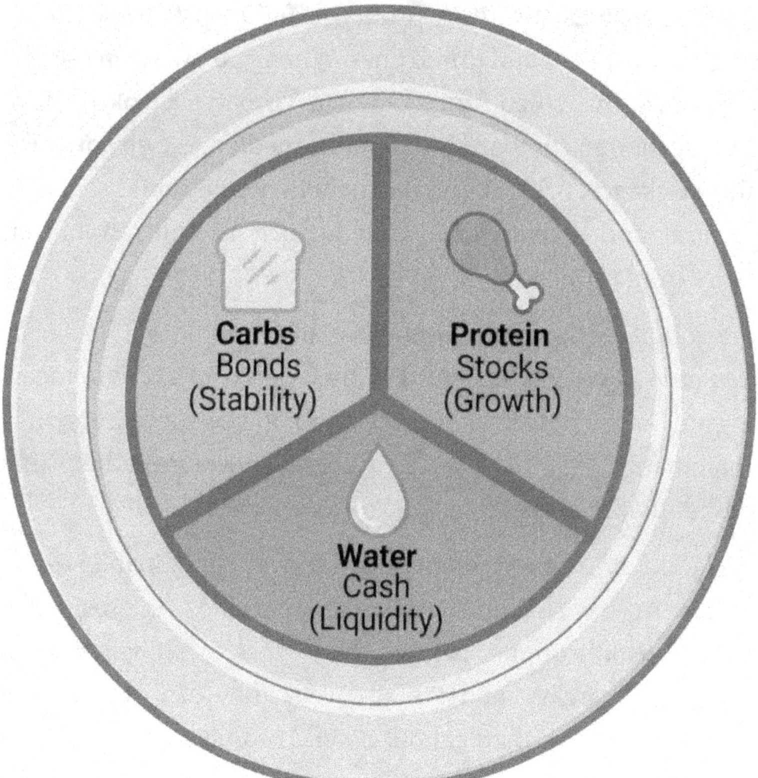

Think of those three slices as a balanced meal. Protein builds. Carbs steady. Water keeps everything moving. You can survive a day eating nothing but protein, or nothing but carbs, but try doing it for years while running around with a mortgage, kids, and jobs that occasionally throw curveballs. You'll feel it.

Stocks: the growth engine (and the stomach-drop ingredient)

Stocks are ownership. When you own a stock, you own a sliver of a real business-its future profits, its failures, its competition, its good decisions and dumb ones. When you own *many* stocks through a diversified fund, you own a broad slice of the business economy: companies that make medicines, sell software, ship packages, drill oil, run payment networks, build semiconductors, and invent things your household will rely on ten years from now.

The job of stocks in a household plan is simple: *outgrow inflation over long periods.* If your money needs to become meaningfully more powerful over 10, 20, or 30 years—retirement is the obvious example—stocks are usually the ingredient that makes that possible.

They do it in the most annoying way imaginable. Stock prices are emotional. They react to earnings, interest rates, wars, breakthroughs, rumors, and vibes. That volatility is why they're scary, and it's also why they can offer higher expected returns than calmer assets. People want to be paid for enduring the rollercoaster.

A useful mental reset: a stock fund isn't a savings account with mood swings. It's a long-term claim on business growth, delivered through a pricing mechanism that panics and celebrates in public.

The household mistake with stocks is almost never "stocks exist." It's *using stocks for money that has a short deadline*. If the roof might need replacing next year, a stock fund is a bad roof fund. Not because stocks are "bad," but because their best feature—long-term growth—doesn't help you when the timeline is short and the price could be down right when the contractor wants a deposit.

Bonds: the shock absorbers (and the misunderstood ingredient)

Bonds are loans. When you buy a bond, you're lending money to a government or a company in exchange for interest. When you own a bond fund, you own a basket of those loans.

In household terms, bonds often have two jobs:

- *Reduce the violence of the ride.* Bonds tend to be less volatile than stocks, so they can make a portfolio feel more livable.
- *Provide a source of funds that is less likely to be down at the same time as stocks.* Not guaranteed, but often helpful.

If stocks are the growth engine, bonds are the seat belts and airbags. They don't make the car fast; they make it survivable.

Bonds are also the asset class where misconceptions breed. People hear "bonds are safe" and assume "bonds can't lose money." Bonds *can* lose money—especially in the short run—because bond prices move when interest rates change.

Here's the plain-language mechanics. New bonds come out all the time. If new bonds pay higher interest than old bonds, the old bonds become less attractive, so their market price drops. That's not a moral failing; it's basic competition. The longer a bond's maturity (how long until it pays back principal), the more its price tends to react to interest-rate changes. That's *interest-rate risk*.

There's also *credit risk*: the possibility the borrower can't pay. U.S. Treasury bonds have very low credit risk; a shaky company's bonds have more.

The household implication is surprisingly practical: bonds are about *stability*, but you still have to choose the type of stability you need. Shorter-term, higher-quality bonds usually bounce around less than longer-term, lower-quality bonds. The label "bond fund" is not a personality guarantee.

Bonds often shine in the exact moments households need steadiness: when you're approaching a goal date, when you're transitioning from saving to spending, or when market swings are provoking the kind of fear that turns temporary volatility into permanent mistakes.

Cash: the liquidity reserve (and the ingredient people underestimate)

Cash is not just bills in a drawer. In a household plan, "cash" usually means money that is readily accessible and designed to hold its value in nominal dollars: high-yield savings accounts, money market accounts, money market mutual funds, and short-term Treasury bills (depending on where you hold them and your comfort with the mechanics).

Cash has one job that no other asset class does as reliably: *be there on the exact day you need it*. Not "probably," not "historically," not "over long periods." Tuesday. Rent day. Deductible day. The day a transmission dies. The day a job loss turns into a gap.

Cash is the ingredient that keeps your household from turning market volatility into a forced sale. It buys time, and time is what makes every other part of your plan work.

Cash also has a cost: inflation. If prices rise and your cash doesn't keep up, your purchasing power leaks away. It's quiet, which makes it dangerous. With stocks, you feel risk as a punch. With cash, you feel it as a slow dimming of the lights.

So the goal isn't "cash is bad" or "cash is good." The goal is "cash is *for* something." Emergency funds. Known near-term purchases. Sleep-at-night money. A buffer that prevents you from raiding long-term investments when life gets loud.

A kitchen-table case study: three goals, one paycheck hiccup

Take a household that looks like millions of real ones: Sam and Alex, two incomes, one kid, one aging car, and a calendar full of expensive milestones.

They have three goals:

- Build a six-month emergency cushion because one job is commission-heavy.
- Save for a home down payment in about three years.
- Invest for retirement, 25+ years away.

A well-meaning friend tells them, "Put it all in the market. The market always goes up long-term." That sentence contains a truth and a trap. The truth is the "long-term" part. The trap is the "all" part.

Now picture a rough year. Sam's commission income drops sharply. At the same time, the stock market is down. If their emergency fund and down payment fund are also in stocks, they're squeezed from both sides. They may have to sell investments at depressed prices to cover ordinary life. In other words, volatility becomes permanent loss.

But if they match ingredients to jobs, the story changes. Their emergency fund sits in cash-like holdings: boring, accessible, ready. Their down payment savings might be mostly cash and high-quality short-term bonds-still conservative because the deadline is close. Their retirement contributions go into diversified stock funds (often mixed with bonds, depending on their personal fit), because that money's job is decades-long growth.

Same market. Same household. Very different outcome, because the plan is built around *time horizon* and *liquidity needs.* When the commission dip hits, they use cash. They don't have to touch retirement. They don't have to turn a scary year into a permanent wound.

This is the sneaky power of asset classes: they're not just about returns. They're about *preventing bad decisions under stress.*

The "jobs" framework: what each ingredient is good at (and terrible at)

If you want a household-friendly way to remember the roles, ask one question before you invest a dollar:

> *What job is this dollar supposed to do, and when will I need it?*

Then match the ingredient:

Stocks are good at growing purchasing power over long periods. They are terrible at being predictable next month.

Bonds are good at smoothing the ride and acting as ballast in a portfolio. They are terrible as a "can't lose" promise, especially over short windows when rates are changing.

Cash is good at immediacy and certainty of nominal value. It is terrible at long-term growth because inflation is always running in the background like an app you can't delete.

Households get into trouble when they ask an ingredient to do a job it wasn't designed for—like asking water to build muscle or asking bread to hydrate you. You can try, but you'll be disappointed and confused, and then you'll blame the wrong thing.

A word on "safe" and "risky" (and why those words cause fights)

Couples often get stuck in an argument that sounds like a values conflict but is really a vocabulary problem:

- One person says, "We should be safer."
- The other person hears, "We should give up growth."

The solution is to replace "safe/risky" with more precise language:

- *How soon do we need this money?* (Timeline)
- *How much can we tolerate seeing it drop?* (Emotional tolerance)
- *How much flexibility do we have if life hits us?* (Household resilience)

Now you're not debating personalities. You're designing a system.

It also helps to name the two kinds of danger:

- *Market danger:* prices can fall.
- *Plan danger:* a fall can force the wrong move.

Stocks carry more market danger. Cash carries more inflation danger. Bonds carry a blend that depends on what kind you own. "Safety" isn't a single dial; it's choosing which dangers you can afford and which would break your plan.

Practical guardrails for using each ingredient well

A few grounded rules keep households out of the most common traps:

Don't invest "deadline money" in volatile assets. If you'll need the money on a specific date (or within a narrow window), use cash and/or high-quality short-term bonds. The goal isn't maximum return; it's showing up with the money intact.

Use bonds to stabilize the household, not to win an argument. Some people use bonds as a compromise asset: "Fine, we'll add bonds so you stop worrying." That's backwards. Bonds should be there because they make the whole plan more durable-because they reduce the odds you panic-sell during a stock downturn or because you're closer to spending the money.

Treat cash like a tool, not a verdict. Holding cash isn't "giving up." It's paying for flexibility. The question is whether you're buying the right amount of flexibility for your life.

Actionable takeaway: build a one-page "money menu"

This week, take 30 minutes and make a one-page menu for your household. Draw three rows labeled *Cash* (*Liquidity*), *Bonds* (*Stability*), *Stocks* (*Growth*). Then list your real goals on the right side with a date next to each one (emergency fund: always; car replacement: 18 months; down payment: 3 years; retirement: 25 years; sabbatical: 5 years). For each goal, choose the ingredient that matches the job.

If you disagree, don't argue about "risk." Argue about timelines and consequences: "If this money drops 25% right before we need it, what happens in our life?" The right allocation is the one that lets you keep your long-term money working *and* keeps your short-term life from hijacking the plan. That's a balanced meal you can actually live on.

6.3 The Power of Diversification

The most expensive sentence in household investing is also one of the most ordinary: "I'm pretty sure this one's a winner."

It might be a coworker's company right before an IPO. It might be a trendy sector everyone's talking about. It might be your spouse's employer stock because it feels loyal, or because it's gone up so much that selling feels like stepping off a moving train. The story is always the same: the brain meets a good narrative and starts mistaking familiarity for safety.

Diversification is the antidote, and it's almost comically unglamorous. No secret. No hot take. Just the decision to stop trying to be right about one narrow thing—and instead own a wide slice of *everything*. In finance, that boring move has a nickname: the only free lunch.

"Free lunch" doesn't mean you get higher returns without tradeoffs. It means you can often reduce risk without necessarily reducing expected return, simply by avoiding avoidable, concentrated bets. You're not paid extra for taking risks that can be easily diversified away. That kind of risk has a name: *idiosyncratic risk*—the risk of one company, one industry, one country, one "theme" blowing up. You can keep the broader market risk (the economy can have bad years) while trimming the "my one pick was wrong" risk.

A quick gut-check: if the success of your plan depends on one CEO, one product launch, one court case, or one regulatory change, you're not investing—you're auditioning for stress.

Correlation: the secret ingredient behind the free lunch

Diversification works because not everything moves in sync. Some assets zig while others zag. The degree to which two things move together is called *correlation*. You don't need the math to use it. Think of correlation as a friendship dynamic:

- Perfectly correlated friends show up to every party together and leave together. Great company, but if they cancel, you're alone.
- Uncorrelated friends are the ones who can't both be busy at the same time. One has your back when the other disappears.

In markets, correlation is the same idea: when one part of your portfolio is having a bad day, you want another part that's less likely to be having the exact same bad day *for the exact same reason*.

To make that intuitive, picture two small businesses on the same street.

One sells umbrellas. The other sells ice cream.

Their profits don't rise and fall together. Rainy days are great for the umbrella shop and awful for the ice cream shop. Sunny days flip the script. Neither business is "better." They simply thrive under different weather.

That mismatch is the magic.

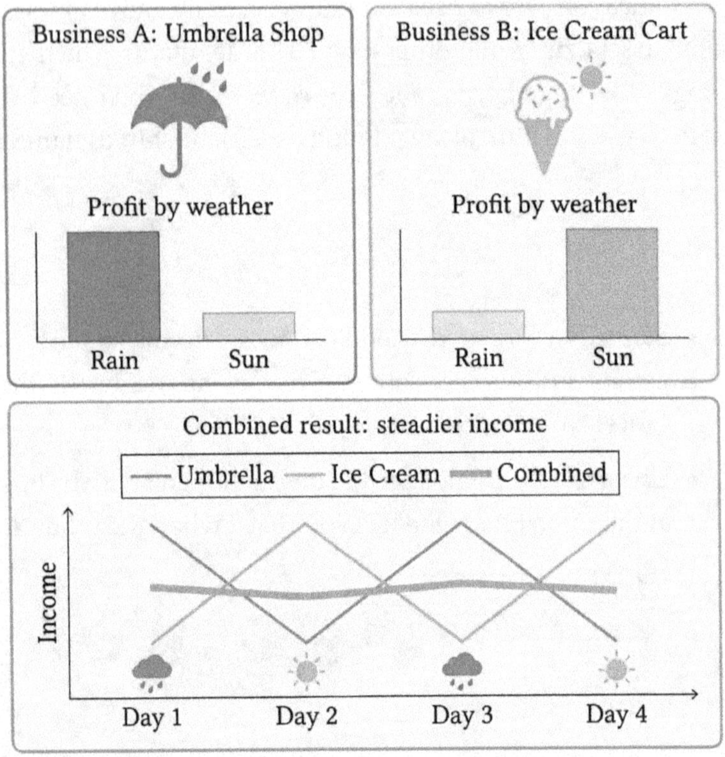

The combined income line is smoother not because either business became "less risky," but because their ups and downs don't line up. That's correlation in one glance.

Your portfolio can work the same way. If you own a variety of things that respond differently to the same world, your financial life becomes less hostage to any single kind of weather.

Owning everything vs. picking winners: a household reality check

Picking winners is appealing because it feels like control. It's also seductive because it occasionally works—and when it works, it rewards you with a story you can tell at dinner. "We bought *that* stock early." You don't get a fun story from owning a total market index fund. You get quiet progress and a balanced nervous system.

The problem with "winner picking" in a household plan isn't that it's always wrong. It's that it creates *fragility*. One big bet can dominate your outcome, and that's the opposite of what most two-income households need. Your life already has concentrated risks: your city's job market, your health, your housing costs, your kids' needs, your parents' aging. A portfolio that adds more single-point failure isn't bold—it's redundant.

Concentration risk shows up in sneaky ways:

- Employer stock: your paycheck and your portfolio depend on the same company.
- Sector obsession: your investments rise and fall with one industry's cycle.
- "Home country only" investing: your portfolio leans entirely on one economy and one currency.

None of these feel reckless in the moment because they're familiar. Familiar is comforting. It's also a terrible risk metric.

Diversification is how you make sure one bad break doesn't get to rewrite your whole plan.

A concrete case study: the "safe" blue-chip that wasn't

If you want a cautionary tale that doesn't involve a meme stock or a crypto exchange, consider something that used to be shorthand for safety: General Electric.

For decades, GE was a symbol of American industrial strength. It was a "blue chip," a stalwart, a stock people held for the long run, often as a cornerstone position. Yet GE's stock experienced a long, painful decline from its peak around 2000, and it also slashed its dividend during the financial crisis. Many households who treated it like a permanent, reliable anchor learned a brutal lesson: a single company—even an iconic one—can stumble for years. Sometimes the business changes. Sometimes management makes damaging decisions. Sometimes the world shifts under it.

The point isn't "GE is bad." The point is that *one-company risk is uncompensated.* If you can get broad exposure to the economy through diversified funds, why bet your future on the internal decisions of one boardroom?

This is also why "I'll just buy the biggest companies" is not the same as diversification. Big companies can still be concentrated in the same forces—interest rates, energy prices, consumer demand, regulation, currency swings. Diversification is about *many sources of return,* not just many tickers that all sneeze together.

Diversification doesn't prevent losses; it prevents disasters

A diversified portfolio can absolutely go down. In a bad year for the overall stock market, most stocks will be down at the same time. That is *market risk,* and you can't diversify it away if you want stock-market returns. Diversification isn't a force field. It's a disaster-reduction system.

It reduces the odds that:

- one company scandal ruins your retirement,

- one industry downturn wipes out your down payment,
- one country's political mess derails your future,
- your household is hit by a "double whammy" where income and investments break in the same place.

That last one is why diversification matters so much for two-income households. Your human capital—your ability to earn—is already exposed to the economy. If your investments are also narrowly exposed, you've stacked the same risk twice.

The practical version: diversification by default

Most households don't need to build diversification by buying 47 individual stocks and 23 bonds. That's not a plan; that's a hobby with a tax bill.

Practical diversification usually means broad, low-cost funds that already own lots of underlying holdings. A total U.S. stock fund, a total international stock fund, and a broad bond fund can give you thousands of securities in a few building blocks. That's "own everything" in a way that fits into a busy life.

If you want one guiding principle: diversify across *companies*, *sectors*, and *geographies*. Your grocery bill doesn't care whether the S&P 500 had a great quarter. Your kid's braces don't care whether tech is hot. The more your portfolio relies on one narrow slice of the world, the more random your outcomes become.

Rebalancing: the quiet discipline that keeps risk from drifting

Diversification is not a one-time event. Over time, your portfolio will drift because different assets grow at different speeds.

After a strong stock run, stocks become a bigger slice of your holdings. After a stock crash, they become a smaller slice. If you do nothing, you don't just get different returns—you end up taking a different level of risk than you intended.

Rebalancing is the act of bringing your portfolio back to its target mix. It's simple, but psychologically weird: it often asks you to do the opposite of your instincts.

- When stocks have surged and everyone feels brilliant, rebalancing nudges you to sell some winners.
- When stocks have cratered and everyone feels nauseated, rebalancing nudges you to buy what's on sale.

This is why rebalancing is so valuable for households. It turns a vague intention ("we're long-term investors") into a mechanical behavior that quietly enforces "buy low, sell high" without requiring you to predict anything.

A concrete example:

Suppose you and your partner choose a 70%/30% mix: 70% stocks, 30% bonds. After a strong bull market, it might drift to 80/20. That means you're now taking more stock risk than you agreed to—whether you realize it or not.

Rebalancing means trimming stocks and adding to bonds to get back to 70/30. You're not doing it because you think a crash is imminent. You're doing it because your household chose a risk level that you can live with, and you're refusing to let a hot streak silently rewrite the deal.

After a big downturn, the reverse happens. Your 70/30 might become 60/40. Rebalancing asks you to buy stocks when they feel dangerous. That's not a personality test; it's a risk-control system. It's also one of the cleanest ways to avoid the classic mistake of selling after losses and buying after gains.

Rebalancing rules that real households can follow

Households don't fail because they don't know what rebalancing is. They fail because they make it too complicated, too frequent, or too emotional.

A few rules that work in the real world:

- Rebalance on a schedule (often once or twice a year) *or* when your allocation drifts beyond a set band (for example, 5 percentage points). Not both. Keep it boring.
- Use new contributions when possible. If you're adding money regularly, you can often rebalance by directing new dollars to the underweight asset instead of selling anything.
- Don't rebalance based on vibes. The point is to avoid "headline allocation," where your portfolio becomes a reflection of this week's fear.

If you're investing through workplace plans, rebalancing can be as simple as adjusting where new contributions go and occasionally exchanging between funds inside the account. The magic is not the tool—it's the discipline.

The emotional benefit: diversification reduces the number of reasons to panic

Household investing is rarely derailed by one big, dramatic decision. It's derailed by a series of small, exhausting ones: checking, worrying, tweaking, chasing, regretting.

Diversification reduces how many things you have to be right about. You no longer need to know which company wins the AI race, which bank avoids a bad loan book, which retailer nails the next trend, which country handles inflation best. You're not betting on a single horse; you own the racetrack.

That's not just a financial advantage. It's a relationship advantage. Fewer concentrated bets means fewer nights of one partner saying, "Are we sure about this?" and the other partner pretending to be sure. A diversified plan is easier to hold together when life gets messy, because it doesn't require heroic confidence.

Actionable takeaway: write a "no single point of failure" rule

This week, open your accounts and look for any holding that could single-handedly derail your plan. Then write one household rule and stick it somewhere visible (notes app counts):

> *No single company, sector, or theme gets to decide our future.*

Make it specific by adding a cap you both agree on—something like, "No single stock will be more than 5% of our investable assets," or "Employer stock will be trimmed each quarter until it's below X." Then choose a rebalancing rule (annual calendar date *or* drift bands) and put it on your calendar like a dentist appointment: not fun, not optional, wildly cheaper than ignoring it.

Diversification won't make markets gentle. It makes your household harder to break.

6.4 Investment Vehicles

The ingredient list is only half the story. You can decide you want "stocks and bonds" the way you can decide you want "vegetables and protein," but you still have to choose how you'll actually buy, store, and serve them without turning dinner into a second job.

Investment vehicles are the containers: the wrappers that hold the ingredients. Most households don't get into trouble because they picked the "wrong" asset class; they get into trouble because they bought the right ingredient in a messy container—too expensive, too complicated, or too easy to tinker with when the market gets dramatic.

If your goal is maximum efficiency and minimum hassle, you're looking for vehicles that do three things well:

- give broad exposure (so you're diversified by default),
- keep costs low (because fees compound against you),
- reduce the number of decisions you have to make on a random Tuesday.

Those are the vehicles that fit real life: a household with two jobs, a calendar full of obligations, and limited patience for financial micromanagement.

Index funds: buying the haystack on purpose

An *index fund* is a fund designed to track a market index—basically a pre-made list of investments. Instead of trying to find the "best" companies, it aims to own the broad market (or a broad slice of it) and deliver the market's return, minus a small fee.

The appeal isn't just that index funds are often low-cost. It's that they are *philosophically aligned* with a household trying to keep money boring: you stop pretending you can outguess the collective wisdom (and emotional overreaction) of millions of buyers and sellers, and you take what the market gives.

There's a quiet humility to indexing that is deeply practical. You don't need to be right about which company dominates next decade. You don't need to be right about which sector gets disrupted. You just need the world economy to keep doing what it's done for a very long time: innovate, grow, and occasionally panic.

Index funds are the "self-driving car" of long-term investing. You still have to pick the destination (your goals and timeline) and you still have to keep your hands available (rebalancing, adjusting contributions, not driving into debt), but you're no longer steering every second.

Mutual funds: the bus

A *mutual fund* is a pooled investment. You and many other investors put money into a single pot. The fund uses that pot to buy the underlying assets-stocks, bonds, or both-according to its strategy.

Mutual funds are priced once per day, after the market closes. The price is based on the fund's net asset value (NAV), which is just "what all the holdings are worth" divided by the number of fund shares. When you buy or sell, you get the end-of-day price.

That once-a-day pricing is why mutual funds feel like a bus: you don't get to control the exact moment the doors close. But the bus is predictable. It comes daily. You can set up automatic purchases. You can ignore intraday noise. For a household trying to avoid constant tinkering, that's a feature, not a bug.

Mutual funds often shine inside retirement accounts, especially in workplace plans where they're the default building blocks. They're also friendly for automatic contributions: "every payday, buy $X" is easy, and it's one of the most powerful household behaviors you can automate.

Where mutual funds can get tricky is in the fine print: some have sales loads (front-end or back-end commissions), some have higher operating expenses, and some have share classes that look almost identical but charge different fees. The fund itself isn't the problem; the fee structure can be.

ETFs: the taxi

An *ETF* (exchange-traded fund) is also a pooled investment, but it trades on an exchange like a stock. That means you can buy and sell throughout the day at market prices, which may be slightly above or below the value of the underlying holdings (the NAV). Usually the gap is small for large, liquid ETFs, but it exists.

If mutual funds are buses, ETFs are taxis. You can hop in at 10:17 a.m. or 3:58 p.m. You can place different types of orders. You can move quickly.

For some households, that flexibility is useful-especially in taxable brokerage accounts where ETFs can be convenient and, in some cases, more tax-efficient because of how creations and redemptions work in the background. (That back-end plumbing is handled by institutions called authorized participants, who create or redeem ETF shares in large blocks. You don't need to manage it; you just benefit from the structure when it helps.)

For other households, the taxi is a temptation. Intraday trading turns investing into a button you can press when you're bored, scared, or overconfident. ETFs make it *possible* to trade constantly. They don't make it *wise*. If you know you're the kind of person who will "just check" and then "just adjust," the convenience can become an expensive personality tax.

Index fund vs. mutual fund vs. ETF: the naming trap

People often talk as if these are three separate categories, but they overlap.

An index fund can be structured as a mutual fund *or* an ETF. "Index" describes *how* the fund invests (tracking an index). "Mutual fund" and "ETF" describe *how* you buy and sell it (the wrapper).

That's why you'll see a low-cost "total market index" offered in both a mutual fund share class and an ETF share class at some providers. Same ingredients, similar recipe, different container.

If you want fewer headaches, focus on what matters:

- What does it own? (Broad and diversified, or narrow and concentrated?)
- What does it cost? (Expense ratio and any other fees.)
- How will I behave with it? (Will the wrapper tempt me to meddle?)

The real-world case study: the "simple" portfolio that got complicated

Consider a household you might recognize: Danielle and Marco, mid-30s, two incomes, one kid, another possibly on the way. They're responsible people. They decided to "get serious" about investing after a friend posted impressive gains.

They opened a brokerage account and started buying ETFs because it felt modern and efficient. At first, it was a broad U.S. stock ETF and a bond ETF—great start. Then they discovered themed ETFs: clean energy, cybersecurity, "disruptive innovation," and a dividend ETF that promised "income." Each purchase felt like adding a smart ingredient. Each one had a persuasive story.

Within a year, they owned 14 ETFs.

Here's what went wrong, and it had nothing to do with intelligence:

- They accidentally doubled up on the same big companies across multiple ETFs.
- They created a portfolio that was *harder* to rebalance because it had too many moving parts.

- They checked prices more often because ETFs trade like stocks, and each ticker became a new source of emotional input.
- They paid higher average fees than they realized because "just a little more expensive" repeated 14 times adds up.

The moment that snapped them back to reality wasn't a spreadsheet. It was a bad market month. Their phone became a slot machine: refresh, drop, refresh, drop. They started having the same argument in different clothes: "Should we sell the risky ones?" "Which ones are the risky ones?" "Why do we have so many?"

They weren't failing at investing. Their vehicle choices had quietly turned investing into a hobby, and hobbies don't mix well with anxiety.

Their fix was almost disappointingly simple: they consolidated into a small set of broad, low-cost index funds and set automatic contributions. They didn't become less ambitious. They became less distractible.

That's the vehicle lesson in one household story: the best wrapper is the one that helps you *keep* your plan when your feelings are loud.

Target-date funds: the "set it and mostly forget it" option

A *target-date fund* is a single fund that holds a diversified mix of stocks and bonds and gradually shifts to a more conservative allocation as the target year approaches (often a retirement year). It's designed to be an all-in-one solution.

For busy households, target-date funds can be a gift: instant diversification, automatic rebalancing, and a built-in glide path that reduces risk over time. In a workplace retirement plan, they're often the cleanest way to avoid decision fatigue.

The catch is that "target date" doesn't magically know *your* life. Two people retiring in the same year can have wildly different pensions, savings rates, health situations, and willingness to work longer. The fund's glide path is a reasonable default, not a personalized plan.

There's also an account-type nuance: if you hold a target-date fund in a taxable brokerage account, you may face taxable distributions depending on the fund's activity. In retirement accounts (401(k), 403(b), IRA), that's usually less of a concern because the account itself shelters the tax friction.

Still, for many households, target-date funds hit the sweet spot: low maintenance, good enough, and behaviorally protective.

What "efficient" really means: costs, taxes, and friction

Households often define efficiency as "the fund went up a lot." That's performance, not efficiency.

Efficiency is what you *keep* after the invisible leaks:

- *Expense ratio:* the annual cost of running the fund, expressed as a percentage.
- *Trading costs:* bid-ask spreads for ETFs, commissions (if any), and the friction of frequent trading.

- *Taxes:* especially in taxable accounts, where dividends and capital gains distributions can create a bill even if you didn't sell.

A household plan is like a bathtub you're trying to fill for decades. Returns are the faucet. Fees and taxes are tiny holes in the tub. A tiny hole doesn't look like much—until you realize it leaks *every year*.

That's why low-cost, broad index vehicles are so powerful. They don't just reduce expenses. They reduce the need for constant decisions, which reduces the odds you do something costly at exactly the wrong time.

A practical selection checklist that avoids 90% of the nonsense

When you're staring at a list of funds in a 401(k) or browsing ETFs in a brokerage account, it's easy to get hypnotized by names. Names are marketing. Your household needs filters.

A short checklist that works:

- *Does it do a simple job?* Broad U.S. stocks, broad international stocks, broad bonds. If the job description sounds like a headline, be cautious.
- *Is it diversified by design?* The fewer holdings, the more you're betting on specific outcomes.
- *Is the expense ratio low?* Low doesn't mean "free," but it usually means you're not voluntarily compounding a drag.

- *Will this wrapper make us fidget?* If your partner is already nervous about volatility, an intraday-trading wrapper might not be the kindest choice.

Notice what's not on the list: last year's return. Past performance is a loud storyteller, and it loves tricking people into buying what already had its moment.

Household-friendly defaults (because willpower is a limited resource)

If you want maximum efficiency and minimum hassle, a few default approaches tend to work well:

- In workplace retirement accounts: a low-cost target-date fund *or* a simple mix of broad index mutual funds.
- In IRAs: low-cost index mutual funds or ETFs, chosen for broad exposure and low expenses.
- In taxable brokerage accounts: broad, low-cost index ETFs or tax-efficient index mutual funds, with minimal trading.

The pattern is consistent: broad, low-cost, easy to automate, hard to overmanage.

That last part—hard to overmanage—is not a joke. Many households don't need more financial options. They need fewer ways to sabotage themselves.

Actionable takeaway: pick your "forever containers" and lock in the boring

This week, do a container audit. For every account you have (401(k), IRA, brokerage), write down the vehicle you're using and answer one question: *Does this wrapper make it easier or harder to stick to our plan?*

Then pick one "forever container" for each account type—something broad, low-cost, and boring enough that you won't feel the urge to tinker. Set automatic contributions where you can. If you're using ETFs, add one extra guardrail: no trading during market hours unless it's scheduled rebalancing day. Investing doesn't need more adrenaline; it needs more consistency.

6.5 Costs, Fees, and Tax Drag

If investing is the engine that helps a household outrun inflation, then fees are the sand in the gas tank. Not dramatic, not cinematic—just relentlessly grinding away at your future while you"re busy living your life.

That"s what makes fees so dangerous: they don"t feel like a loss. There"s no moment where someone hands you a receipt that says, "Congratulations, you just paid $38,417 for the privilege of owning your own money." The cost arrives invisibly, shaved off a little every year, compounding against you the same way returns compound for you. It"s the financial equivalent of a slow leak in a tire: you can still drive—until you can"t.

Taxes can do the same thing. Not the obvious "I sold something and paid capital gains" kind (though that matters). The sneakier kind is *tax drag*: the steady reduction in growth because taxes are triggered earlier or more often than they need to be. Two families can own the same underlying investments and end up with different results because one family"s setup keeps more of the compounding inside the account.

The good news is that costs and tax drag are among the few parts of investing you can actually control. Markets will do what they do. Your job is to stop paying unnecessary tolls.

The simplest fee you'll ever pay-and the one that matters most

Most households meet fees through a number called an *expense ratio*. It's the annual cost a fund charges to operate, expressed as a percentage of the money you have invested in that fund. A 0.05% expense ratio means you pay about $5 per year for every $10,000 invested. A 1.00% expense ratio means you pay $100 per year for every $10,000 invested.

Those numbers look small because they are presented as percentages, and humans are famously bad at feeling percentages. But your portfolio doesn't experience fees as a one-time subtraction. It experiences fees as a permanent headwind.

A one-time $1,000 expense hurts. A permanent 1% drag hurts *forever*—including on the growth you would have earned on that money. Fees don't just take dollars; they take *future dollars*.

A concrete lifetime-feel calculation: what a 1% fee can cost

Let's put some household-scale numbers on this.

Assume:

- Starting balance: $50,000
- Annual contribution: $10,000 (made over the year; we'll model it annually for simplicity)
- Gross market return (before fees): 7% per year
- Time horizon: 30 years

Now compare two scenarios:

- Low-cost investing: 0.05% annual fund cost
- High-cost investing: 1.05% annual fund cost (that extra 1% could be higher fund expenses, an advisory fee, a wrap fee, or a combination)

Net returns become roughly 6.95% vs. 5.95%. One percentage point doesn't sound like a life-altering difference. Over 30 years, it can be.

Below is a visual that makes the "fee gap" impossible to unsee.

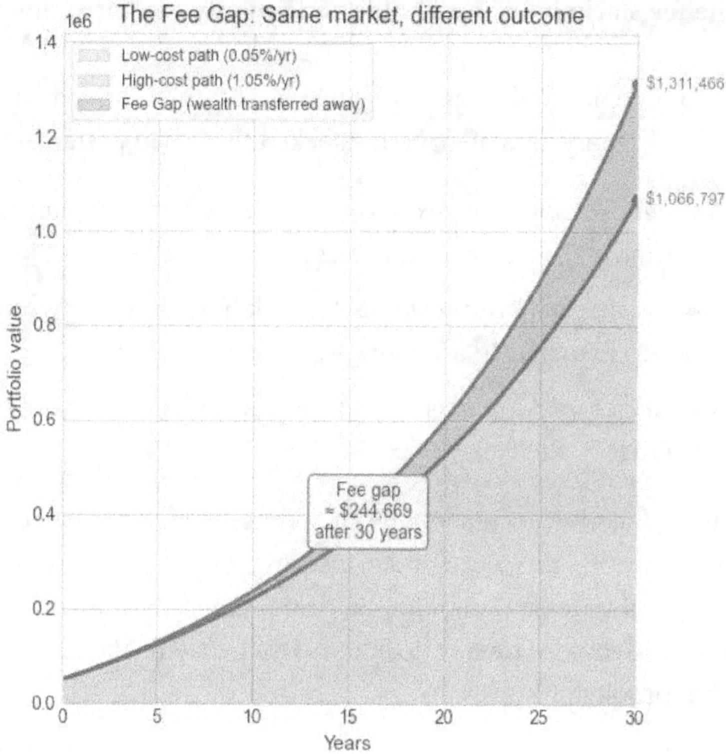

This chart is why fees are called the termites of your financial house. You don't notice the damage at first. Then one day you look up and realize the "missing" money isn't missing at all-it quietly became someone else's revenue.

If you want the emotional translation: that red gap can represent years of earlier retirement, the ability to work less, the freedom to absorb a layoff without panic, or the difference between your kids taking loans versus taking opportunities.

The fee stack: how households accidentally pay three times

Many people think they're paying *one* fee: the fund expense ratio. In reality, fees often come stacked like a bad parfait:

- *Fund-level fees*: the expense ratio (and, for some mutual funds, 12b-1 distribution fees).
- *Account-level fees*: custodial fees, platform fees, or "wrap" fees in managed accounts.
- *Advice fees*: a percentage of assets under management (AUM), flat fees, or hourly planning fees.

Paying for advice is not inherently wrong. A good planner can prevent costly mistakes: under-insuring, taking on the wrong debt, failing to save enough, or panic-selling in a downturn. But households deserve to know what they're paying and what problem it solves.

A useful line to keep your household honest: "If this fee disappeared tomorrow, what would break?" If the answer is "nothing," you just found an unnecessary leak.

A real-world case study: the 401(k) that looked fine until it didn't

Here's a scenario that plays out in workplaces everywhere.

Nina contributes diligently to her 401(k). She picks a fund option with a respectable-sounding name—something like "Growth Opportunities" or "Strategic Allocation"—and never thinks about it again. The plan website is clunky, so she doesn't log in often. Her contributions increase slowly over time, and the balance grows. Life is busy; the account is on autopilot.

Years later, she switches jobs and rolls the money into an IRA. That's when she sees the old fund's expense ratio: 1.20%. She also learns the fund had additional trading costs and a history of distributing taxable capital gains in comparable taxable accounts (not her issue in a 401(k), but a sign of turnover and friction). Nina didn't do anything reckless. She just let the default choices in a plan quietly choose her fee level.

The painful part isn't that she paid fees. It's that she paid *high* fees without receiving anything she would have knowingly bought:

- No meaningful simplification (the fund didn't reduce decision-making; it just charged more).
- No tailored planning (the fee didn't include retirement modeling for her household).
- No unique access (it wasn't some rare asset class she needed).

That's the most common fee problem: not paying for value, but paying by inertia.

Taxes: the other invisible leak

Taxes deserve the same respect as fees because they also compound—just in reverse.

In the U.S., the biggest levers households bump into are:

- *Account type*: tax-advantaged (401(k), IRA, Roth) vs. taxable brokerage.
- *Turnover and distributions*: some funds generate more taxable events than others.
- *Dividends and interest*: different investments throw off different kinds of taxable income.

Tax drag is less about "taxes are bad" and more about timing. The earlier taxes are triggered, the less money remains inside the account to compound. Two investors can earn the same pre-tax return, and the one with better tax timing ends up with more wealth simply because more dollars stayed invested longer.

This is why "high turnover" strategies can be quietly expensive in taxable accounts. If a fund buys and sells frequently, it can realize gains inside the fund and pass them through to shareholders as distributions—creating taxes even if you never sold your fund shares. That can feel like being handed a bill for a party you didn't attend.

ETFs often get attention here because their structure can, in many cases, reduce capital gain distributions compared to mutual funds. That's not a guarantee, and it's not universal. The practical household point is simpler: in taxable accounts, prefer vehicles and strategies that don't create unnecessary taxable churn.

Trading friction: the "I didn't know I was paying that" cost

Some costs don't show up as a neat line item called "fee." They show up as:

- bid-ask spreads (especially in less liquid ETFs),
- commissions (less common now, but not gone everywhere),
- the performance drag of frequent trading,
- the tax consequences of short-term gains.

Frequent trading is often sold as sophistication. For households, it's usually just a way to pay more—more friction, more taxes, more emotional energy—without a reliable increase in returns.

If your investing app makes trading feel like ordering takeout, that's not convenience. That's a business model.

How to plug the leaks without making investing your hobby

Households don't need perfection. They need a few high-impact decisions that stay true year after year.

1) Treat expense ratio like an ingredient label, not a trivia fact. When comparing similar funds (same broad asset class, similar exposure), a lower expense ratio is one of the most reliable advantages you can buy. Not because it makes markets better, but because it guarantees you keep more of whatever markets deliver.

2) Pay for advice intentionally-or don't. If you're paying an advisor a percentage of assets, ask what you're getting that justifies a fee that rises as your balance rises. Sometimes it's worth it. Sometimes a flat-fee planning arrangement fits better. Sometimes the best "advisor" is a simple plan you can follow without needing a referee.

3) Put the right investments in the right accounts when you can. As a broad rule of thumb, tax-advantaged accounts are precious real estate. Use them for long-term investing. Use taxable accounts thoughtfully, aiming for tax-efficient holdings and low turnover.

4) Avoid paying "complexity premiums." If a product is hard to explain, it's often hard to monitor—and complexity is where fees hide. Households benefit from vehicles that are transparent enough to understand at the kitchen table.

Actionable takeaway: a 20-minute "leak audit" you can do tonight

Open every investing account you have and write down:

- each fund's expense ratio,
- any account or advisory fee you pay (percentage or dollars),
- whether the account is taxable or tax-advantaged.

Then do one decisive thing: pick one high-fee holding you can replace with a broadly diversified, lower-cost alternative *in the same asset class*, and set a date to make the change (or to ask your plan provider how). If you can't change it—some workplace plans have limited menus—at least know the cost, and offset it elsewhere by keeping everything you can control lean.

The goal isn't to win a spreadsheet contest. It's to stop quietly donating pieces of your future to fees and avoidable taxes, so the compounding you're working for actually stays in your household.

Chapter 7

Building a Resilient Portfolio

If your household income is the engine of your escape plan, your portfolio is the transmission that converts that power into forward motion. Too often, families stall out because they treat investing like gambling-chasing hot tips or panicking when the news gets scary. In this chapter, we stop trying to outsmart the market and start building a fortress around your money. We will design a strategy that accounts for the biggest threat to your wealth (taxes) and the most unpredictable variable in finance (your own brain).

7.1 The Family Financial Constitution

Picture a normal Tuesday evening in a two-income house. Someone's heating leftovers. Someone else is trying to pay a bill from their phone while a kid asks for help with homework. Then the notification hits: *Market sinks on recession fears.* Or: *Tech stock soars 18% after earnings.* Or the more personal version: a coworker at lunch says, "I moved everything to cash. You should too."

In that moment, you are not deciding about a portfolio. You are deciding about *fear, pride, regret,* and (if you're partnered) a quiet power struggle: *Who's right about money? Who gets blamed if this goes wrong?*

A household Investment Policy Statement (IPS) is the antidote. It's a short, written agreement that answers the questions you don't want to improvise in a panic: What are we investing for? How much risk are we choosing on purpose? What are the rules when the market gets weird? It is not a fancy institutional document. It's a family constitution: a set of laws you draft in peacetime so you don't start arguing in a storm.

Most couples already have an IPS. It's just unwritten, inconsistent, and heavily influenced by the last scary headline. One spouse believes in *"safe"* (meaning cash). The other believes in *"growth"* (meaning whatever has been going up). You might even agree in theory, but your actions reveal the real policy: maybe you chase performance, sell when it hurts, and buy when it feels comfortable again. That *is* a strategy. It's just the kind that quietly taxes your future.

A written IPS turns the strategy from a mood into a system.

Here's the uncomfortable truth: in a dual-income household, the stakes of emotional investing are higher than people admit. Two incomes usually mean higher fixed costs: a larger mortgage approved on two paychecks, childcare contracts you can't cancel, car payments tied to commutes, and a lifestyle that silently assumes "both of us working" is permanent. That's the classic two-income trap: the margin for error shrinks. When the market drops, the decision isn't only "Should we sell?" It becomes "Can we still handle our life if one of us gets laid off *and* our investments are down?"

The IPS won't prevent layoffs or bear markets. It prevents a third disaster: *self-inflicted damage.*

A good IPS does two jobs at once.

First, it protects your portfolio from your emotions. That's the obvious part: no impulsive selling, no revenge-trading, no *"all-in on the hot thing."*

Second, it protects your relationship from your portfolio. That's the underrated part. Money fights often aren't about money; they're about *trust*. When the market is down, every decision carries an implied accusation: "You made us do this." A written constitution changes the emotional physics. Instead of spouse vs. spouse, it becomes both of you vs. the problem, following the rules you agreed to when you were calm.

The biggest misconception is that an IPS is only for "serious" investors. It's actually most valuable for households who *don't want* investing to become a second job. If your life is already full—deadlines, carpools, aging parents, a calendar that looks like a crime scene—you need investing to be boring. Boring is not neglect. Boring is *repeatable*. Boring is what lets compound growth do its quiet work while you live your actual life.

So what goes into a family financial constitution?

Not ten pages. Not legal language. Not a dissertation on economic cycles. You want something you can read in under five minutes and still feel your shoulders drop.

A practical IPS usually fits on one or two pages and answers six core questions:

1. *What is this money for, and when do we need it?*

2. *How will we measure success (and what doesn't matter)?*
3. *What do we invest in, and what do we refuse to invest in?*
4. *What is our target mix (stocks/bonds/cash or similar), and what are the acceptable ranges?*
5. *When do we rebalance, and who is responsible for pushing the button?*
6. *What events justify changing the plan?*

Those questions look simple on paper. In real life, each one defuses a common household money bomb.

Start with purpose and timeline, because this is where otherwise-smart couples accidentally sabotage themselves. "We're investing for the future" is not a usable sentence. The future has dates.

Try this instead: list three goals, each with a time horizon.

- *Short (0–3 years):* emergency reserves, near-term home repairs, the next car replacement.
- *Medium (3–10 years):* down payment upgrade, possible one-income sabbatical, private school decision, starting a business.
- *Long (10+ years):* retirement, financial independence, legacy giving.

Then draw a bright, almost obnoxious line between *investing money* and *sleep-at-night money*. Stocks can be great long-term wealth engines, but they are terrible short-term savings accounts. A family IPS doesn't just say what you'll invest in; it says what you *won't gamble with*. That sentence alone prevents the classic mistake of putting next year's needs into this year's market.

Next, define success in a way that doesn't trick you into self-sabotage.

Most people unconsciously define success as "beating the market" or "making the right calls." That turns your portfolio into a test of intelligence. And when money becomes a test of intelligence, it becomes a test of ego. Ego is expensive.

A calmer definition of success sounds like:

> *We consistently invest $X per month, keep costs low, stay diversified, and maintain an allocation that lets us keep investing through downturns.*

Notice what's missing: predictions. A family IPS is a refusal to audition for a role as "the one who saw it coming."

Now the part couples tend to skip because it feels technical: what you will invest in, and what you will avoid.

This is not about finding the perfect funds. This is about preventing "portfolio sprawl," where you accumulate random holdings like souvenirs: a few old 401(k) choices, a handful of stocks a friend mentioned, an inherited mutual fund with a high fee, and a crypto position you're afraid to admit you don't understand anymore. Portfolio sprawl creates the illusion of diversification while actually creating confusion—and confusion is fertile ground for impulsive decisions.

Your IPS should name your allowed building blocks in plain English. For many households, that means broad, low-cost index funds (funds that simply aim to match a market index rather than trying to pick winners). Your constitution might say something like:

> *We will use diversified stock index funds and high-quality bond funds as our core holdings. We will not buy individual stocks as "investments" (only as a limited hobby allocation). We will not buy products we cannot explain to each other in two minutes.*

That "two-minute explanation" rule is a marriage-saver. It's also a protection against the modern financial marketplace, which is very good at selling complexity as sophistication.

Then comes the heart of the IPS: your target asset allocation.

Asset allocation is just your mix of different types of investments—commonly stocks and bonds. Stocks are volatile but historically have offered higher long-term growth potential. Bonds tend to be steadier but usually grow more slowly. The mix you choose is not a math problem; it's a behavior problem. The "best" allocation is the one you can actually stick with when the news turns apocalyptic.

This is where two-income households need unusual honesty. It's easy to claim you have a high risk tolerance when both paychecks are flowing and the market is calm. It's harder to stick with that claim during a layoff scare, a medical bill, or a sudden childcare crisis. A realistic IPS doesn't ask, "How brave are we?" It asks, "What allocation would still let us sleep, work, and parent without checking our balances every hour?"

A useful trick: instead of picking one number, pick a target and a guardrail. Example:

> *Target: 80% stock / 20% bonds. Acceptable range: stocks may drift between 75% and 85% before we rebalance.*

That range is not a loophole. It's a way to avoid frantic micromanagement. If you rebalance every time the market twitches, you create constant decisions, and constant decisions create decision fatigue. Decision fatigue is when your brain, tired of choosing all day, starts choosing based on emotion and convenience. In other words: exactly how people blow up their portfolios.

Rebalancing is the constitution's enforcement mechanism.

Rebalancing means you periodically bring your portfolio back to its target mix. If stocks rise a lot, they become a bigger percentage of your portfolio; rebalancing would mean selling a bit of what went up and buying what lagged. If stocks fall a lot, they become a smaller percentage; rebalancing would mean buying stocks when they feel scary.

That's the entire point: rebalancing forces you to behave like the person you *wish* you were. The IPS turns "buy low, sell high" from a motivational poster into a calendar reminder.

There are two household-friendly rebalancing rules:

- *Calendar rule:* review and rebalance once or twice a year (often enough to stay on track, rare enough to stay sane).
- *Threshold rule:* rebalance only when your allocation drifts beyond a preset band (like 5 percentage points).

Pick one. Put it in writing. Then decide who is responsible. Not who "cares more about investing," but who will reliably do it.

The IPS should name the operator and the back-up operator, because life happens.

Now for the part that makes the IPS feel less like a finance document and more like a marriage document: *change control.*

Change control is a fancy term for a simple idea: what events are allowed to change the plan?

Without this, "We should change our strategy" becomes a constant argument, because the market provides endless excuses. Every rally makes you feel late. Every drop makes you feel smart for selling. Every hot sector makes you feel behind. Every scary headline makes you feel irresponsible for staying invested.

Your IPS should name the handful of legitimate reasons to change course. For example:

- a major change in household income (job loss, new job, one spouse leaving work)
- a major change in timeline (planning to buy a home in 18 months)
- a major change in liquidity needs (new child, caregiving responsibilities)
- a major change in risk capacity (large debt added, business volatility)

And then include a sentence that sounds almost too strict—until you need it:

> *We will not change our long-term allocation based on market forecasts, social media, or news headlines.*

That one line blocks a thousand bad decisions.

A real-world example makes this concrete.

Consider Maya and Chris (a composite of many real households). They're both in their late 30s with two kids. Maya works in marketing at a mid-size company; Chris is a project manager in commercial construction. For years, they did what lots of competent adults do: they contributed to their 401(k)s, kept some money in savings, and occasionally bought a stock someone mentioned. They were "good with money," meaning bills got paid and they didn't carry a credit card balance.

Then 2022 happened, and their accounts fell. They didn't lose their jobs, but they lost their confidence. At dinner, Chris would say, "Maybe we should pause the 401(k) for a while and rebuild cash." Maya would say, "That's when you're supposed to keep investing." The conversation would loop: fear vs. logic, protection vs. optimism, both people feeling unheard.

The problem wasn't intelligence. The problem was that they were trying to negotiate a constitution during a crisis.

So they wrote one. Not a manifesto—a page.

They agreed on three things that changed everything:

1. Their emergency fund target: 6 months of essential expenses (because construction can be cyclical).
2. Their long-term allocation: 75/25 stock/bond, using diversified funds, with a 5-point rebalance band.
3. Their rule for downturns: contributions continue automatically unless a job loss forces a temporary pause; if the market falls sharply, they do not "wait for clarity" be-

fore investing, because clarity usually arrives after prices rebound.

Then they added an unusual clause: any proposed change requires a 72-hour waiting period and a written reason tied to one of their change-control events. That meant no more midnight "sell everything" conversations after a scary podcast. If it still felt necessary three days later *and* it matched a legitimate trigger, they could discuss it calmly.

The next time the market dropped, they still felt the emotional punch. They just didn't turn it into a marital negotiation. The IPS didn't make them fearless. It made them *consistent*.

A family IPS also gives you a place to handle the question couples are often afraid to say out loud: *How much freedom does each person get?*

Some households do best with full joint decisions. Others do best with clear boundaries. A constitution can include a "sandbox" allocation, sometimes called a hobby account: a small percentage (say 0–5%) for speculative bets or individual stock picking. The key is that the sandbox is sized so that losing it would be annoying, not life-altering, and it's explicitly separated from the core plan.

That solves two problems at once: it scratches the itch to tinker without endangering the family's future, and it reduces the chance that one spouse feels financially policed.

Here's another clause that belongs in more households than you'd think: *communication rules*.

Money conversations tend to happen when one person is anxious and the other is tired. The IPS can specify a monthly or quarterly "money meeting" with a short agenda: check cash flow, confirm contributions, review allocation drift only if it crosses the band, and note upcoming expenses. If you don't schedule money, money schedules you—usually at 10:47 p.m. when you're both at your worst.

Keep the meeting short. Use a timer if you have to. The goal isn't to turn your marriage into a finance committee. The goal is to reduce financial ambiguity, because ambiguity breeds anxiety.

The constitution should also name the enemies you're specifically defending against. Write them down like a wanted poster:

- *Performance chasing:* buying what just went up because it feels safe.
- *Panic selling:* selling what went down because it feels dangerous.
- *Story addiction:* believing a compelling narrative is the same as an investment thesis.
- *Over-monitoring:* checking balances so often that normal volatility feels like an emergency.

When you name a behavior, you can build a rule around it. For example: "We check balances on the first Saturday of the month only," or "We do not watch financial news while the kids are awake," or "We do not place trades from a phone."

That last one sounds silly until you realize how many expensive decisions happen in the grocery store parking lot.

If you're wondering how formal to make this, aim for "signed and boring." Literally sign it. Put the date. Put it somewhere you can find it. If you use a shared cloud folder for household documents, save it there. If you're analog, print it and put it with your tax file. The point is to treat it like what it is: a governing document.

A useful add-on is an "If we are tempted to break the rules" paragraph. It might say:

> *If we feel an urgent need to make a change because of market news, we will (1) wait 72 hours, (2) reread this IPS, (3) verify whether a change-control trigger has occurred, and (4) make any changes only during our scheduled money meeting.*

That paragraph is a seatbelt. You don't wear it because you plan to crash. You wear it because you're human.

One more reality check: an IPS does not guarantee you'll pick the best possible portfolio. It guarantees you'll avoid the most common ways ordinary families destroy wealth: selling low, buying high, and letting a stressful season of life turn into permanent financial damage.

In a two-income household, that's a superpower. Your edge is not secret information. Your edge is *staying invested* while life is happening.

If you want a simple template, here's a one-page constitution you can adapt. Write it in your own voice. The best IPS sounds like your household, not like a bank.

Our Household Investment Policy Statement (Date: ________)
Purpose: We are investing to (1) retire around ________, (2) fund __________, and (3) maintain flexibility for __________.
Time horizon: Long-term (10+ years) for retirement assets. Short-term goals will be held in cash or low-volatility accounts.
Target allocation: ________% stocks / ________% bonds (range: stocks ______–______%).
Allowed investments: diversified, low-cost index funds/ETFs; high-quality bond funds; cash equivalents for reserves.
Disallowed investments (unless in sandbox): individual stocks, leveraged products, anything we can't explain clearly, and "hot tips."
Rebalancing rule: (choose one) annual / or when allocation drifts beyond _____ percentage points.
Monitoring cadence: portfolio check on ________ (e.g., first Saturday monthly).
Change-control triggers: job loss, major income change, major timeline change, major liquidity need.
Decision rule in stress: no same-day changes based on headlines; 72-hour waiting period.
Signatures:______________________ ______________________

Actionable takeaway: open a shared document tonight and draft your IPS in plain language, then schedule a 30-minute meeting within the next week to finalize the target allocation, rebalancing rule, and change-control triggers—and sign it like you mean it.

7.2 Why Your Brain is Your Portfolio's Worst Enemy

The market doesn't usually ruin families with one dramatic punch. It does it with a thousand tiny, very human reactions: checking an account balance at the wrong moment, doom-scrolling a headline written to spike adrenaline, texting your spouse "Should we do something?" while standing in line at Target. The portfolio becomes a mood ring. And moods make terrible fund managers.

The problem isn't that you feel fear when prices fall or excitement when they rise. The problem is that your brain treats those feelings like *instructions*.

A resilient investing plan can be simple. Sticking to it is the hard part. Not because you're irrational or weak, but because your brain is doing exactly what it evolved to do: keep you alive in a world where "danger" used to mean teeth and claws, not a red number on a screen.

Start with the single most expensive wiring glitch in personal finance: *loss aversion*. In plain language, losing money hurts more than gaining the same amount feels good. A $1,000 drop feels like an insult; a $1,000 gain feels like a pleasant surprise you're not sure you deserve. That imbalance sounds like a minor quirk until you watch it in the wild: people hold onto losing investments too long (to avoid admitting defeat) and sell winners too soon (to "lock it in"), or they bail out entirely after a drop and then demand "proof" it's safe before getting back in.

The market doesn't ask for proof. It offers prices. Your brain asks for comfort. Comfort is usually most available near market highs, which is exactly when future returns tend to be least generous.

Two-income households are especially vulnerable to this because your risk isn't only market risk. It's life risk: job volatility, childcare, health, elder care, and the very real possibility that the two incomes you planned around won't always arrive on schedule. When you already feel stretched, a market decline doesn't feel like volatility; it feels like the floor shifting under the entire house.

That's why the most dangerous sentence in investing isn't "I think the market will go down." It's "We can't afford for it to go down."

If you hear yourself thinking that, the problem is rarely the portfolio. It's that you're asking the portfolio to do the emergency fund's job, the insurance policy's job, and the marriage's job (emotional reassurance) all at once.

Then comes *recency bias*: the brain's tendency to treat what just happened as what will keep happening. After a year of strong returns, your mind quietly edits the future to match the recent past. After a sharp decline, it edits the future toward continued pain. You can see this bias everywhere outside finance. If you've had a string of rough nights with a newborn, it feels like you will never sleep again. If you've had a few smooth months at work, you start assuming the calm will last.

Markets are the same. They seduce you with patterns and then punish you for believing them too literally.

Recency bias is how investors end up buying what's already expensive and selling what's already cheap, while fully believing they are being prudent. They aren't trying to "buy high, sell low." They're trying to stop feeling stupid. They're trying to stop feeling scared. The market doesn't care what you're trying to feel.

There's also *overconfidence*, which has nothing to do with intelligence and everything to do with being human. Overconfidence shows up as "I can time this," "I can spot the next big thing," or the classic: "I'll sell now and buy back when it settles down." That last one feels like a neat trick until you realize it requires two perfect decisions: when to get out and when to get back in. Most people struggle with one.

Overconfidence is sneaky because it often arrives wearing the costume of responsibility. "I'm just being careful," you tell yourself, while you hit the button that locks in a loss.

And then there's *herding*: the pull to do what other people are doing simply because other people are doing it. When your coworker is bragging about a hot stock, it doesn't feel like gossip. It feels like you're being excluded from a wealth opportunity. When your neighbor says they moved everything to cash, it doesn't feel like fear. It feels like they know something you don't.

Herding is social pain disguised as financial insight.

Put these biases together and you get a predictable emotional cycle. It's not just "up is happy, down is sad." It's more personal than that: up makes you feel *competent* and down makes you feel *exposed*. Competence leads to risk-taking at exactly the wrong time. Exposure leads to retreat at exactly the wrong time.

The tragedy is that the cycle is self-reinforcing. Buy at a high and you are more likely to experience regret. Regret makes you more likely to sell at a low. Selling at a low makes you more likely to sit out the rebound. Sitting out the rebound makes you more likely to chase later. It's a treadmill made of feelings.

The only reliable way to win against this treadmill is to treat your brain like a well-meaning but unqualified advisor. You don't fire it. You just stop letting it trade.

That's what guardrails are: rules that reduce the number of emotional decisions you're forced to make.

Before the guardrails, it helps to see the cycle clearly—especially the moment where families do the most damage: panic selling. Panic selling isn't only about money; it's a craving for relief. You sell because you want the pain to stop. The market is offering you the emotional equivalent of ripping off a bandage. It feels better immediately. It can leave a scar for decades.

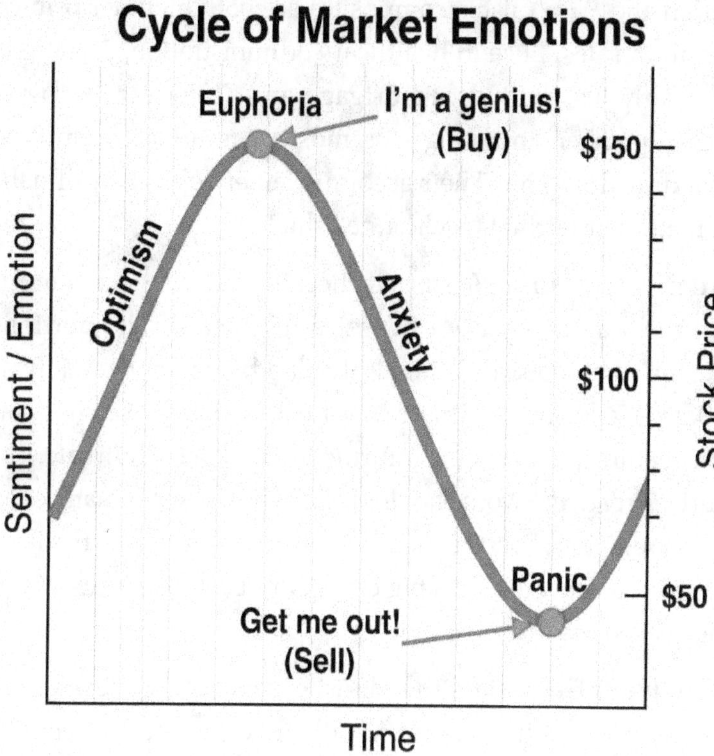

A chart like that can feel almost insulting because it's so simple. And yet it matches what happens in real kitchens, real group chats, real marriages. The line goes up and down, but our confidence goes up and down *harder*. We feel smartest near the peak, when risk is highest. We feel most desperate near the bottom, when expected long-term returns are typically more attractive. The math doesn't change; the feelings do.

Here's a concrete case study that shows how it plays out in a two-income life.

Jordan and Elise (also a composite) are both high performers: Jordan in sales, Elise in healthcare administration. They're in their early 40s, two kids, a mortgage, and a calendar that never shuts up. Their investing was mostly on autopilot until an ugly drawdown hit. Their accounts fell enough that it finally pierced their normal "we're fine" shield.

Jordan reacted first. He started checking balances every morning, the way he checked his sales dashboard. The portfolio became a scoreboard. When it dropped, he felt personally attacked. He began reading market commentary at night, convincing himself that being informed would make him safe. The more he read, the worse he slept. Then he did what many capable people do when they feel powerless: he grabbed the lever he could reach. He sold a big chunk to cash, intending to "buy back when the coast is clear."

Elise didn't feel calm either, but her anxiety expressed itself differently: she didn't want to talk about it. She had enough stress. The market felt like one more demand. Their money conversations became short and sharp. Jordan thought Elise was being reckless by not engaging. Elise thought Jordan was being reckless by making moves while wired with fear.

Weeks later, the market rebounded. Jordan felt relief, then embarrassment. He didn't buy back in immediately because now he needed certainty. But certainty had a price: higher prices. He waited. Prices rose more. His "responsible" move turned into a slow-motion trap: he was out of the market during the bounce, and now re-entry felt like admitting a mistake.

What fixed it wasn't a better forecast. It was a better *process*. They made two changes:

- They reduced monitoring. Checking quarterly felt almost negligent at first, like ignoring a baby monitor. But within a month, Jordan's stress fell sharply, because he stopped injecting market noise into his nervous system daily.
- They created pre-decided actions for scary moments. If markets dropped beyond a certain point, their only allowed action was to verify their cash reserves and keep contributions going. Selling required a waiting period and a joint decision in daylight.

The surprise wasn't that the portfolio recovered (markets do what they do). The surprise was that their relationship recovered faster. They stopped turning volatility into a referendum on each other's competence.

That's the core lesson: the investing mistakes that cost the most money also cost the most *peace*. And peace has compounding returns, too.

So what guardrails actually work in the real world, when you're busy and tired?

One of the most powerful is simply controlling the frequency of evaluation. Your brain interprets frequent checking as frequent danger. If you look every day, you will experience a stream of losses even in a rising market, because daily noise includes plenty of down days. Your mind tallies those down days as threats. Then it starts to demand action.

A household-friendly rule is: check contributions monthly (to confirm money is flowing), check allocation quarterly or semiannually (to rebalance if needed), and check goal progress annually (to adjust savings rate, not to "outsmart" markets). You're not ignoring reality; you're refusing to let random short-term fluctuations rent space in your head.

Another guardrail: a *pre-commitment script* for the moment when fear spikes. This matters because panic makes you forget what you know. You can be financially literate and still, in a bad moment, act like a different person.

Write a script you can actually say out loud:

> *We expected declines. This is the cost of long-term growth. Our job is not to predict. Our job is to follow the plan: verify cash reserves, keep investing, rebalance only by rule.*

It sounds corny. It works anyway. Under stress, your brain craves a simple sentence to hold onto.

A third guardrail is structural: make the right behavior the default.

When investing is manual, you invest when you feel optimistic. When investing is automatic, you invest when you're busy, when you're bored, when you're worried, when you're living your life. Automatic contributions are a way of quietly buying through every emotional season. They force you to do the one thing that matters more than cleverness: keep showing up.

Another structural guardrail is separation of roles. Many couples function best when one person is the "operator" (handles the mechanics) and the other is the "auditor" (reviews and signs off periodically). This reduces friction and reduces the temptation for both people to intervene every time the market sneezes.

The operator doesn't get to freelance. The auditor doesn't get to micromanage. The constitution is the boss.

Then there's the "fun money" guardrail: a small sandbox for experimentation. This is not permission to gamble with the rent; it's a pressure-release valve. When someone in the household has a strong itch to pick stocks, chase themes, or try the latest investing app, a sandbox keeps that itch from invading the entire plan.

A good sandbox has three rules:

- It is capped (a percentage you can lose without changing your life).
- Losses stay in the sandbox (no topping it up to "get back to even").
- The core plan is not touched to fund it.

This isn't about shaming curiosity. It's about quarantining it.

You also need a guardrail against "I heard a thing." Markets are information-rich and wisdom-poor. A useful household rule is the *two-minute test*: if you can't explain the investment to your spouse in two minutes, without jargon, you don't buy it. Not because complexity is always bad, but because complexity is often a hiding place for hope.

Hope is not a strategy. It's an emotion. Emotions belong in your life, not in your asset allocation.

One more guardrail that sounds small but is huge in practice: ban trading when you're dysregulated. If you're angry, exhausted, or scared, you don't place trades. Your brain is in "fight or flight," which is excellent for avoiding car accidents and terrible for deciding whether to sell an index fund.

A simple household policy: no trades after 8 p.m., no trades on a phone, no trades on days when either spouse says "I'm not okay." Make the rule strict enough that you can't wiggle through it when you're in a spiral.

Behavioral finance can sound like a lecture about all the ways humans are flawed. That's not the point. The point is compassion with teeth: designing a system that assumes you will be stressed sometimes, busy often, and emotional occasionally—and still gets the job done.

Financial success is less about being the calmest person in the room and more about being the person whose plan doesn't require constant calm.

There's a final twist worth naming, because it's so common in high-achieving couples: you may be using investing as a place to prove yourself. The market becomes a stage where you get to be the smart one, the responsible one, the one who "handles money." That identity can feel good until it becomes fragile. Then every decline feels like a threat to your self-image, and the urge to act becomes overwhelming.

If you recognize that in yourself, it's not a character flaw. It's a cue. Move the goal from "be right" to "be consistent." Consistency is less glamorous than being right. It's also far more profitable.

Actionable takeaway: pick two guardrails and implement them this week—(1) set a fixed schedule for checking investments (monthly for contributions, quarterly for allocation), and (2) write a one-paragraph panic script plus a hard rule that major changes require a 72-hour pause and a joint, daylight decision. Then keep the script somewhere you can reach with shaking hands.

7.3 It's Not What You Earn, It's What You Keep

A weird thing happens when a household starts earning "real money." The paychecks get bigger, but the sense of progress doesn't always follow. You can be doing everything right—working hard, maxing the match, staying out of credit card debt—and still feel like your investing results have a leak.

Often, that leak is taxes.

Not the once-a-year, file-your-return kind of taxes. The quiet, ongoing kind: the taxes that skim a little off dividends, a little off interest, a little off capital gains distributions, a little off trading, and then compound in the wrong direction for twenty years. It's death by a thousand paper cuts, except the paper cuts are coming out of your future.

High-income households (and many two-income households become high-income almost by accident) have a particular problem: your marginal tax rate can be punishingly high, which means every avoidable dollar of taxable investment income is more expensive than it looks. The goal isn't to "beat taxes." The goal is to stop paying more than you owe simply because your money is parked in the wrong place.

The core idea is called *asset location*. That sounds like jargon, but it's simply this: *which investments go in which accounts*. You already know about asset allocation—your mix of stocks and bonds. Asset *location* is the seating chart.

Your household probably has multiple "rooms" where money can live:

- *Tax-deferred accounts* (like a traditional 401(k) or traditional IRA): you generally get a tax break now, but withdrawals in retirement are typically taxed as ordinary income.
- *Tax-free accounts* (like a Roth IRA or Roth 401(k)): you contribute after-tax money, and qualified withdrawals later are generally tax-free.
- *Taxable brokerage accounts*: no special wrapper; taxes show up along the way in the form of dividends, interest, and realized capital gains.

Asset location is the art of putting the tax-inefficient stuff behind the strongest "tax glass," and leaving the more tax-friendly stuff in the open.

If this feels like a small optimization, consider a blunt reality: taxes compound too. Your return isn't what your investments earn. Your return is what you earn *after* fees, *after* taxes, and *after* your own behavior gets a vote. That last part is why households create written rules and guardrails; this part is why the same discipline should extend to where you place your holdings.

Here's the first principle that makes asset location click:

Not all investment income is taxed the same.

Interest from many bond funds is typically taxed as ordinary income in a taxable account. Ordinary income is the same category as your paycheck. For a two-income household, that can be a painful rate.

Long-term capital gains (profits on investments held more than a year, in general) are often taxed at preferential rates. Qualified dividends (certain dividends from U.S. and eligible foreign companies, when you meet holding-period rules) can also receive those preferential rates. Preferential doesn't mean "free," but it can be meaningfully lower than ordinary income.

So in a taxable account, you generally prefer investments that generate *less* taxable income each year and more of their return in the form of long-term appreciation. That usually points you toward broad, tax-efficient stock index funds and away from high-interest, high-turnover strategies.

Principle two:

Some taxes are optional—if you avoid creating the taxable event.

A taxable event is something like selling at a profit (realizing capital gains) or receiving certain distributions. You can't control what the market does, but you can often control whether you trigger taxes by churning your portfolio. This is another place where "activity" is quietly expensive: every time you sell something in a taxable account, you're not just making a bet. You might be mailing a check to the IRS.

Principle three is the one that feels almost unfair:

Fees compound the same way taxes do—relentlessly.

A fund's expense ratio can look like a rounding error. But over decades, small annual drags can create huge differences in ending wealth because you're not just losing the fee; you're losing the growth on the fee, year after year. This matters in every account type, but it becomes especially irritating when you're also paying taxes in a taxable account. You don't want "fee drag" and "tax drag" working as a tag team.

So what does a sensible household seating chart look like?

A simple, practical rule-of-thumb is:

- Put *tax-inefficient* assets in tax-advantaged accounts.
- Put *tax-efficient* assets in taxable accounts.
- Put your *highest expected growth* assets in Roth-style accounts (because you want that growth to be sheltered).

That's the theory. Here's the translation.

Tax-inefficient assets commonly include:

- taxable bond funds (because interest is generally taxed as ordinary income in taxable accounts)
- REIT funds (real estate investment trusts often distribute income that may be taxed less favorably than qualified dividends)
- actively managed funds with high turnover (because they may distribute capital gains you didn't choose)

Tax-efficient assets commonly include:

- broad stock index funds and ETFs designed to minimize taxable distributions
- municipal bond funds (for certain households in higher brackets, because the interest is often exempt from federal income tax—though the math depends on yields and your tax situation)

And the "highest expected growth in Roth" idea often means: if you're holding stocks and bonds across multiple accounts, you may prefer that more of the stock exposure ends up in Roth space, while bonds live more in traditional 401(k)/IRA space—*when the overall household allocation still matches the constitution you've written.*

A lot of couples get stuck right here because they think asset location requires a perfectly engineered portfolio. It doesn't. It requires two decisions:

1. What is the household's target allocation?
2. Given the accounts we have, where is the least-bad place to hold each asset?

"Least-bad" is a useful phrase in household finance. You're coordinating old employer plans, current employer plans, maybe a rollover IRA, maybe an HSA, maybe a taxable brokerage. You don't need perfection. You need a system that consistently avoids the biggest unforced errors.

A concrete example makes this real.

Nina and Omar are married, both 37, with a toddler and a baby. Nina works in tech; Omar is an attorney. They're not flashy, but they're high earners. They contribute aggressively to retirement accounts and also invest in a taxable brokerage because they want the option to buy a larger home in 7—10 years (and they don't want every dollar locked up until retirement).

Their portfolio looked "diversified" at first glance: a handful of funds in each account. But the placement was backwards:

- In the taxable brokerage, they held a bond fund and a high-turnover active stock fund.
- In the traditional 401(k), they held a broad stock index fund.
- In Roth IRAs, they held conservative balanced funds.

Nothing about that arrangement was immoral or reckless. It was just expensive.

Every year, the bond fund in taxable threw off interest taxed like salary. The active fund generated distributions they didn't control. Meanwhile, their Roth accounts—the most valuable tax shelter in the house—were stuck doing low-growth work.

They didn't need a new investment philosophy. They needed a seating chart.

They reorganized so that:

- Bonds moved primarily into the traditional 401(k).
- The highest-growth stock funds moved into Roth IRAs.

- The taxable brokerage held a tax-efficient broad stock index fund (and, for their risk comfort, a modest amount of municipal bonds instead of taxable bonds).

Result: their *allocation* stayed essentially the same, but the tax drag dropped. More of their taxable account's return came from long-term appreciation rather than annual ordinary income. Their Roth space finally did what Roth space is designed to do: shelter growth.

And the best part? Their system became easier to maintain. Fewer surprise distributions. Fewer "why did we owe so much this year?" conversations. Less friction in the marriage.

Asset location also forces one of the most grown-up conversations in household finance: *what do we want taxable money for?*

Taxable accounts are not automatically "bad." They're flexible. They can fund early retirement, a sabbatical, a move, a down payment, or simply give you options. For households trying to escape the two-income trap, flexibility is not a luxury. It's a pressure-release valve.

But because taxable accounts create taxes along the way, you want them to be *intentionally designed*. A taxable brokerage account full of interest-heavy holdings is like buying flexibility and then paying rent on it every year.

Two-income households also need to be aware of the *Net Investment Income Tax* (*NIIT*). NIIT is an additional 3.8% tax that can apply to certain investment income once your modified adjusted gross income exceeds certain thresholds (commonly cited thresholds are $200,000 single and $250,000 married filing jointly). The details matter and your tax professional can confirm what applies to you, but the practical takeaway is simple: if your household is near or above those lines, the "keep" rate on investment income can be meaningfully worse than you assume.

That changes the math on decisions that generate taxable income: holding interest-heavy assets in taxable, realizing gains unnecessarily, or churning the portfolio. When people say "taxes matter more when you make more," NIIT is one reason why.

Capital gains deserve their own moment, because this is where households accidentally donate money to the government out of impatience.

In general, if you hold an investment for more than a year before selling, you may qualify for long-term capital gains treatment rather than short-term. Short-term gains are typically taxed like ordinary income. Long-term gains often get preferential rates.

This turns "waiting" into an investing skill. Not waiting for the perfect price—waiting to avoid turning what could have been a preferentially taxed gain into a wage-like gain.

If you want a concrete reference point, the IRS publishes long-term capital gain bracket thresholds by tax year. For example, for the 2025 tax year, the 0% long-term capital gains bracket went up to $96,700 for married filing jointly (and $48,350 single), and the 15% bracket extended up to $600,050 for married filing jointly (and $533,400 single), with amounts above that generally at 20% (before considering NIIT and other factors). The exact numbers change over time, but the behavior lesson does not: short-term gains are often expensive; long-term gains can be less so.

Qualified dividends are another "sounds boring, saves money" detail. Some dividends qualify for preferential tax rates, but only if you meet specific requirements, including holding-period rules. Many households inadvertently reduce tax efficiency by hopping in and out of funds or stocks around dividend dates. The dividend itself can feel like "free money," but it's often just a slice of the investment's value paid out in cash—and in a taxable account, that slice may come with a tax bill.

A dividend is not a bonus. It's a form of return, and the IRS cares about its category.

Then there's the wash-sale rule: the rule that punishes you for trying to claim a tax loss while keeping essentially the same investment.

If you sell a security at a loss and buy a "substantially identical" security within 30 days before or after the sale, the loss is generally disallowed for current tax purposes and is instead added to the cost basis of the replacement shares (again, details matter; the general idea is what counts). This rule exists to prevent people from manufacturing paper losses while not really changing their economic position.

In household terms, wash sales are easy to trigger accidentally:

- You sell an index fund at a loss in a taxable account for tax-loss harvesting.

- Meanwhile, your spouse's retirement account is automatically buying the same fund every payday.

Now your "smart tax move" becomes a mess.

This is one reason a household financial constitution matters. The tax strategy has to be coordinated across accounts and across spouses. Otherwise you're playing chess with half the pieces.

So what should a dual-income household do with all this, without turning their life into a tax seminar?

Make the system visible. Put your accounts and their intended roles in writing. A simple table can do more than a dozen conversations.

Account type	*What it's best for*	*Common fits (examples)*
Tax-free (Roth)	Sheltering long-term growth; keeping future withdrawals more flexible	Broad stock index funds; higher-growth equity funds (within your risk plan)
Tax-deferred (Traditional 401(k)/IRA)	Sheltering tax-inefficient income; deferring taxes until withdrawal	Bond funds; TIPS funds; REIT funds; balanced funds if choices are limited
Taxable brokerage	Flexibility; goals before retirement; tax-efficient compounding	Tax-efficient stock index funds/ETFs; municipal bonds (if appropriate); cash equivalents for near-term needs
HSA (if eligible)	Triple tax advantage when used correctly; long-horizon medical reserve	Often treated like a stealth retirement account: diversified stock funds if you can pay current medical costs out of pocket

This isn't a universal commandment. It's a starting arrangement. Your employer plan might have limited options. Your household might need more taxable liquidity. You might strongly prefer simplicity over incremental optimization. Those are valid choices—as long as they're conscious.

One more nuance that deserves honesty: asset location is not just about taxes. It's also about behavior.

If placing all bonds in a 401(k) makes it harder for you to rebalance calmly, that "optimal" plan can backfire. If putting all stocks in Roth makes one spouse feel like the Roth is "untouchable" and therefore ignored, that can backfire too. The best asset location strategy is the one that your household will maintain for a decade without resentment, confusion, or constant tinkering.

A small tax improvement that you can actually stick with beats a complex "perfect" setup that creates new arguments.

The simplest way to implement asset location without headaches is to think at the household level rather than the account level. Instead of forcing every account to be a perfectly diversified mini-portfolio, you can treat all accounts as one combined portfolio and assign roles.

For example:

- "The 401(k) is the bond engine."
- "The Roth is the growth engine."
- "The taxable account is the flexibility engine."

Now rebalancing becomes moving money according to roles rather than constantly reshuffling every account.

Actionable takeaway: map your household's accounts on one page, label each as *tax-free*, *tax-deferred*, or *taxable*, then list what each currently holds. Circle anything in taxable that generates ordinary-income-heavy return (like many bond funds or high-turnover funds). Over the next two contribution cycles, begin swapping placements—moving the most tax-inefficient holdings into tax-advantaged space and reserving taxable space for tax-efficient funds—until your allocation stays the same but your "keep" rate improves.

7.4 Price vs. Value: Avoiding the Hype

The most expensive sentence in investing is usually said with a grin: *"This time is different."*

It's the sentence that shows up right before people overpay for something they don't understand. It's the sentence that turns a reasonable household into a late-stage participant in somebody else's party. And it's almost always said in the language of inevitability: "Everyone's using it," "It's the future," "You can't miss this," "It's going to change everything."

Sometimes those statements are even true. The future *does* arrive. The problem is that great innovations can still be terrible investments at the price you pay.

That's the first mental gear to install: *a company can be wonderful and still be overpriced.*

Price is what you pay. Value is what you *get.*

Price is a number on a screen, updated every second, bouncing around with the emotional weather. Value is slower. Value is the business underneath: what it sells, what it earns, what it can likely earn in the future, and how risky those earnings are.

If price is the amount on the menu, value is the food that actually arrives at the table. If price is the Zillow listing, value is the house you'll still be living in after the excitement of the offer wears off and the first plumbing issue appears.

Two-income households need this distinction because hype is a time thief and a stress multiplier. You don't have spare hours to babysit a speculative position, and you don't need the extra marital friction that comes from betting the family's future on a story that sounds good in a group chat.

Hype rarely begins with greed. It begins with a reasonable fear: *getting left behind.*

That fear is especially sharp for dual-income households because you're surrounded by signals of competition. Colleagues negotiate. Friends upgrade houses. Someone's spouse works at a company you've heard of. Someone posts a screenshot of gains. Even if you're responsible and skeptical, your brain quietly translates it into a social question: *Are we doing enough? Are we missing something?*

The market has learned to market itself to that insecurity. It doesn't just sell stocks; it sells identity. You're not buying shares-you're buying a role: early adopter, contrarian, future-builder, "the one who saw it."

That's why "price vs. value" isn't an academic concept. It's a defense against being emotionally recruited.

Here's the simplest way to think about value without needing an MBA: a stock is a claim on a stream of future cash flows.

That phrase sounds technical, but it's just a grown-up version of what you already understand. When you buy a rental property, you're buying future rent (minus expenses). When you buy a business, you're buying future profits (minus costs and competition). With stocks, you're buying a small slice of whatever cash the company can generate for its owners over time-either distributed as dividends, used to buy back shares, or reinvested to grow future profits.

A company's *price* can swing wildly without the underlying business changing much at all. A company's *value* can improve for years while the price does nothing. The market is a voting machine in the short run and, more often than people like to admit, a weighing machine in the long run.

So why does hype work so well?

Because stories are easier than spreadsheets. Because "we're changing the world" feels better than "our profit margin is under pressure." Because your brain wants certainty, and hype offers an illusion of certainty in the form of confidence and community.

And because there is a psychological trap called *anchoring*. If you first see a stock at $300 and later it's $180, your brain thinks "cheap," even if $180 still implies a heroic future. The anchor is not value; it's the previous price. In household terms, anchoring is why a sale sign makes you feel smart even if you didn't need the item and the "sale" price is still overpriced.

To avoid hype, you don't need a crystal ball. You need a few blunt questions that pull you back to Earth.

Question 1: What has to go right for this price to make sense?

This is the fastest hype-detector available. If a company is priced like it will dominate its industry, maintain high margins, avoid competitors, and grow rapidly for a decade, then the investment is not "Will it grow?" The investment is "Will it grow *enough to justify this price*?"

Often, the right answer is: it might grow tremendously and still disappoint, because the expectations baked into the price are even more tremendous.

Question 2: What is the simplest explanation for why people are excited?

Sometimes the simple explanation is great: a product is genuinely better, costs are falling, adoption is rising. Sometimes the simple explanation is darker: people are excited because other people are excited. That's reflexive demand-the financial version of a crowded restaurant with a line out the door. The line might mean the food is amazing. It might also mean everyone saw the same TikTok.

When excitement is social, price becomes detached from value faster than you expect.

Question 3: If I am wrong, what happens to my life?

This is the question most hype ignores, and it's the question two-income households can't afford to skip. Speculation becomes dangerous when it can change your timeline, your stress level, or your marriage.

If a bet going wrong means you need to work five extra years, you didn't make an investment-you made a lifestyle commitment without discussing it.

A concrete case study shows how normal people get pulled in.

In 2021, Paul and Renee (composite) were doing what many responsible couples do: contributing to their 401(k)s, building a brokerage account, and trying to keep childcare costs from swallowing their entire raise. Paul's coworkers wouldn't stop talking about a handful of "can't lose" stocks. The charts looked like escalators. The stories were seductive: new technology, new era, "legacy companies are dead."

Paul didn't want to be reckless. He wanted to be *competent*. So he moved a meaningful chunk of their taxable account into a concentrated basket of hyped names. Not all at once-he "scaled in," which felt prudent. Every early gain gave him a hit of relief: proof he wasn't missing the future.

Then the cycle turned. Prices fell. The stories changed. The same people who had preached inevitability now preached patience, then quietly stopped talking about it. Paul felt an urge that is both psychological and deeply human: *get back to even*. That urge is how small mistakes become big ones. He doubled down in the names that had fallen most, not because the value improved, but because the price drop made them feel like bargains.

Renee eventually asked the only question that mattered: "What are these companies worth, and how do we know?"

Paul couldn't answer without repeating the story. That was the moment they realized they didn't have an investment thesis. They had a narrative.

They didn't need to predict the bottom. They needed to reduce the chance that their portfolio became a referendum on Paul's identity. They unwound the concentrated positions gradually, kept a small "sandbox" amount as a reminder (and to avoid the emotional whiplash of a total reversal), and moved the core back into diversified funds aligned with their written rules.

The lesson wasn't "never buy individual stocks." The lesson was that hype makes you confuse a *good story* with a *good price*.

This leads to a subtle point: valuation isn't about precision. It's about *humility*.

You can't calculate a company's exact intrinsic value the way you can calculate the square footage of a house. But you can often tell when the market is pricing something like a fairy tale. A fairy tale price requires a fairy tale future: no recessions, no competition, no regulation, perfect execution, constant growth.

Real businesses live in the real world, where customers change their minds and competitors show up and governments get involved and costs rise. When a stock price assumes the real world won't happen, you're not investing; you're auditioning for a miracle.

One of the most useful instincts in avoiding hype is learning to separate *a great company* from *a great investment*.

A great company might:

- have a product people love
- have a brilliant founder
- have a huge market opportunity

A great investment, at a minimum, requires that you are not overpaying for those qualities.

Households often overpay because they confuse *admiration* with *return*. You can admire a company and still choose not to buy the stock. You can admire a restaurant and still not buy it at ten times its annual profit.

There's another underappreciated trap: *unit bias*. That's the tendency to prefer owning "one whole share" over owning a fraction of a share, or to feel that a lower dollar-per-share price is automatically "cheaper." A $20 stock can be far more expensive (relative to its earnings and prospects) than a $200 stock. Price per share is a labeling decision. Value is a business decision.

Hype loves unit bias. It makes "affordable" sound like "undervalued."

Social media pours gasoline on all of this. Social feeds reward confidence, certainty, and simple narratives; investing rewards patience, humility, and boring repetition. Those incentives are opposites.

There's also a structural problem: many social-sentiment tools are designed to show what people are talking about *now*. Markets price in what people are talking about now extremely quickly. By the time a theme is obvious enough to trend, it's often already in the price. You aren't early-you're well-informed and late.

The antidote isn't to unplug from the internet forever. It's to give hype a place to go that isn't your retirement.

That's where a rules-based approach does its quiet magic. When you buy and sell because your portfolio drifted away from a target allocation, you're not responding to stories.

You're responding to math. Math can be wrong too, but it's at least not trying to impress anyone.

A household constitution that commits you to diversified holdings and rebalancing creates an automatic anti-hype mechanism:

- If a hot sector rockets upward and becomes too large a share of your portfolio, rebalancing forces you to trim it.
- If a hated asset class falls and becomes too small a share, rebalancing forces you to add to it.

That feels emotionally backward, which is exactly why it works.

Still, people want a "valuation shortcut," something they can use without becoming analysts. Here are a few intuition tools that behave well in real family life.

First: *the expectations test*. When a stock is up massively in a short period, the question isn't "Is it a good company?" It's "How much future perfection is already baked in?" Rapid price increases often mean the market has pulled future hopes into the present price. You can still invest, but you should do it with the understanding that you are paying for a lot of good news in advance.

Second: *the competition test*. If the opportunity is truly huge, competitors will arrive. If the company's advantage is not defensible-brand, network effects, switching costs, unique distribution, patents, scale-it's hard to maintain extraordinary profits. Hype often assumes profit without competition, which is like assuming a playground without other kids.

Third: *the cash test*. A business can grow revenue and still be fragile if it constantly needs financing to survive. Cash flow is harder to fake than excitement. You don't need to run a full model, but it's healthy to notice whether the business is actually generating cash or mainly generating headlines.

For two-income households, the most practical way to implement all this is to decide what role individual "hype" investments are allowed to play.

If your constitution already permits a sandbox, hype belongs there. The sandbox is where you can satisfy curiosity without endangering core goals. The moment a sandbox position starts to dominate your net worth, it stops being curiosity and becomes concentration risk wearing a cool outfit.

If you don't have a sandbox and don't want one, that's fine too. The adult move is to say, plainly: "We don't do hype investing with family money." That sentence is a boundary. Boundaries are protective, not restrictive.

There's also a gentle, surprisingly effective household practice: *write down why you bought it*.

Not a long essay. Three sentences:

- What do we believe the business will do?
- What could prove us wrong?
- What would make us sell (other than price movement)?

If you can't write that without referencing the stock price itself, you are not investing-you are reacting.

Finally, remember the quietest point of all: the market will always offer you another exciting story. There is no shortage of "next big things." There is, however, a shortage of time, attention, and emotional bandwidth in a household with two careers and a life to run.

When you choose a boring, diversified approach, you're not opting out of opportunity. You're opting into a different kind of opportunity: the chance to build wealth without turning your evenings into emergency meetings about red candles and rumors.

Actionable takeaway: pick one holding you own (or have been tempted to buy) primarily because of hype, and do the three-sentence test in writing. If you can't describe the business case, the risks, and a non-price-based sell trigger in plain language, move it to a capped sandbox or replace it with a diversified fund that fits your household rules—then set a calendar reminder to stop checking the price for 30 days.

7.5 Stress-Testing Your Financial Fortress

Most households think risk means "the market might go down." That's true, but it's also the *kindest* kind of risk: it's obvious, it's widely discussed, and it shows up in the same place you already look.

The risks that actually wreck plans are quieter. They hide in job benefits packets, in the default settings of a 401(k), in a pile of company shares you never meant to accumulate, in the assumption that you'll always have two paychecks at the same time. They hide in the kind of optimism that sounds responsible: "We'll deal with it if it happens."

Stress-testing is how you stop relying on that optimism.

A stress test is not a prediction. It's a rehearsal. It's you asking, calmly and on purpose: *If a bad year arrives, does our plan still stand up? What breaks first?*

A good stress test starts with an uncomfortable admission: downturns are normal.

Recessions happen. Bear markets happen. Whole sectors go out of fashion. Interest rates move. Real estate can stagnate for years in some areas. None of this requires a villain. It's just the weather of capitalism.

If you build a plan that only works when conditions are sunny, you haven't built a plan. You've built a fairytale.

So the goal isn't to build a portfolio that never feels pain. The goal is to build a household system that can take a hit without forcing you into desperate decisions at the worst possible time.

For a two-income household, stress-testing has a special edge because income is both your strength and your vulnerability. Two paychecks can accelerate savings, but they can also inflate obligations. If the family's "fixed costs" (mortgage, childcare, car payments, subscriptions, student loans) are sized for two incomes, then a job loss doesn't just reduce savings. It turns the monthly budget into a cliff.

In that scenario, the portfolio gets dragged into the emergency. You don't sell because it's a good idea. You sell because the daycare bill is due.

Stress-testing is how you identify those cliffs before you reach them.

Start with the most common hidden fragility: *concentration risk.*

Concentration risk is the danger of having too much of your financial life tied to one company, one sector, one geographic area, or one economic outcome. It's not just about owning a single stock. It's about having many parts of your life that all break in the same kind of storm.

A classic two-income concentration trap looks like this:

- One spouse works for a publicly traded company and holds a large amount of employer stock (through RSUs, ESPP, stock options, or "we never sold it" shares).
- The other spouse works in the same industry or a related one.
- The household also lives in a city whose economy is tied to that industry.

Now a downturn isn't one problem. It's a cascade: layoffs, the local housing market softens, and the portfolio declines *in the exact same direction at the exact same time*. That is not diversification. That is stacking matches next to gasoline.

Employer stock is where concentration risk hides in plain sight, because it arrives dressed as a reward.

RSUs vest. ESPP shares accumulate. Your company's stock has done well historically. You tell yourself you're loyal, or you don't want to pay taxes, or you assume you can "sell later."

Meanwhile, the position quietly grows into a dominant piece of your net worth.

And it feels safe because it's familiar. You know the company's products. You hear internal updates. You recognize the logo. Familiarity is soothing, but it is not protection.

A household stress test treats employer stock as a separate category and asks a blunt question: *If this stock fell 50%, what changes in our life?*

If the answer is "we'd be annoyed," you have a position. If the answer is "we'd delay retirement, cancel plans, or lose sleep for months," you have a vulnerability.

Here's a concrete case study that shows how fast "reward" turns into risk.

Sofia and Mark (composite) were in their mid-30s, both ambitious, no kids yet, saving aggressively. Mark worked at a high-growth tech company and received a steady stream of RSUs. The stock had been a rocket ship. Every vesting date felt like proof that they were doing life correctly. They held nearly all of it, partly because selling felt like disloyalty and partly because the stock's upward slope made selling feel like stepping off a moving escalator.

Then their company hit a rough patch. Guidance changed. The stock dropped. At the same time, Mark's company froze hiring and hinted at layoffs. Sofia's job was stable, but their "future plan" had quietly become "Mark's stock keeps growing."

When they finally did a stress test, they discovered something sobering: over 40% of their investable net worth was tied to

Mark's employer, and their planned down payment for a home was mentally funded by shares that could easily halve again.

They didn't need to hate the company to change the plan. They needed to stop confusing *confidence* with *concentration*.

Their fix was simple and unglamorous:

- They set an employer-stock cap: no more than 10% of investable assets.
- Every vesting event triggered a default sale of the amount above the cap.
- The proceeds were redirected into their diversified core holdings and a clearly labeled "home timeline" fund.

This didn't maximize their potential upside if the stock recovered sharply. It maximized something more valuable for a household: the ability to make life decisions without begging one company's ticker to cooperate.

Stress-testing also means looking for concentration that doesn't look like concentration.

Owning "a bunch of funds" can still be concentrated if those funds are all exposed to the same underlying driver. For example, a portfolio heavy in U.S. large-growth stocks can behave like a single crowded trade during certain periods. A portfolio that looks diversified on the surface can still be heavily tilted toward one style, one region, or one sector.

You don't need to become a quant. You just need a sanity check:

- What percentage is in one company?

- What percentage is in one sector?
- What percentage is tied to one country?
- What percentage is tied to one interest-rate outcome?

If you don't know the answers, you're not necessarily doing anything wrong. You're just driving at night without headlights.

The second major category of household stress-testing is *liquidity risk.*

Liquidity is your ability to access money *without* taking a bad deal. Cash is liquid. A brokerage account is relatively liquid (but may be down when you need it). A 401(k) is generally not liquid without penalties and restrictions. A home is not liquid, even if it is valuable. Your retirement account might say $600,000, but that doesn't mean you can pay a $12,000 emergency with it next week without consequences.

Two-income families get trapped here because they look "wealthy" on paper but are cash-fragile in real life. They have assets, but the assets are locked or volatile, and their monthly obligations are rigid.

A stress test asks: *If one income disappears for six months, what do we pay bills with?*

Not "we'd figure it out." With what money, in what account, in what order?

A practical way to answer is to create a *drawdown order*-a prioritized list of where money comes from in an emergency.

For example, a household might decide:

1. Use the emergency fund (cash).
2. Reduce discretionary spending immediately (because the cheapest dollar is the one you don't spend).
3. Use taxable brokerage funds if needed (mindful of taxes and market conditions).
4. Consider other options only after those (because retirement accounts and home equity have heavier tradeoffs).

The point isn't to script every detail. The point is to prevent the worst-case scenario: being forced to sell long-term assets at depressed prices because the household had no liquidity plan.

Third, stress-testing must include *tax friction*-the way taxes can turn a manageable situation into a messy one.

Taxes matter most when you're under pressure, because pressure forces transactions. If you need cash and the only liquid assets are in a taxable brokerage, selling can trigger capital gains taxes. If the market is down, you might have losses you can harvest, but wash-sale rules can complicate "sell and immediately buy back" behavior-especially if you have automatic purchases elsewhere in the household. If you rebalance aggressively in taxable accounts during volatile periods, you might accidentally create gains at the worst time.

A stress test asks: *In a downturn, do we have a way to raise cash without triggering a surprise tax bill?*

Sometimes the answer is as simple as keeping a slightly larger cash buffer than the spreadsheet suggests. Cash is not "dead money" if it prevents you from making catastrophic, taxed-at-the-worst-time decisions. Cash is insurance against forced selling.

Fourth, stress-testing needs to account for *contribution risk*-what happens to your plan if you can't keep investing at the same pace.

Most households build projections that assume steady contributions: the same percentage of income flowing into retirement accounts every month. Real life rarely cooperates. A recession can bring job loss, reduced bonuses, fewer hours, or unexpected expenses. Even without job loss, stress can make you pause contributions because it feels like control.

A stress test asks: *If we have to reduce contributions for a year, do we still stay on track?*

This is where flexibility becomes a financial asset. If your plan is so tight that missing six months of contributions "ruins retirement," then the plan is fragile. You may need a higher savings rate in good years, a larger emergency fund, or a smaller fixed-cost footprint so you can keep investing even when life is noisy.

Now, the temptation at this point is to ask: "Should we hedge? Should we protect against downturns?"

Hedges can help, but they come with tradeoffs. Insurance costs money, and financial insurance is no different. Many hedging strategies reduce long-term returns or add complexity that households abandon at the wrong time. The best "hedge" for most families is not a clever product. It's a resilient structure:

- diversified holdings (so no single failure breaks you)

- adequate liquidity (so you aren't forced to sell at the bottom)
- reasonable fixed costs (so a job shock is painful but not fatal)
- written rules (so fear doesn't rewrite your plan overnight)

In other words, the fortress is built more with architecture than with weapons.

A household stress test also includes a scenario many couples avoid because it feels almost superstitious: *What if both incomes are hit at once?*

This is not paranoia. It's pattern recognition. Couples often work in correlated industries: two professionals tied to the same economic cycle, or one job that depends on the health of the other (real estate and construction, tech and recruiting, finance and high-end services). During broad downturns, "diversifying incomes" is harder than it sounds.

Ask the uncomfortable questions on purpose:

- If both bonuses disappear, what changes first?
- If one spouse is laid off and the other takes a pay cut, what changes first?
- If a downturn lasts 18 months, what is the plan for health insurance?

These aren't market questions. They're household questions. Your portfolio is only one wall of the fortress.

Stress-testing is also where you confront the difference between *risk tolerance* and *risk capacity*.

Risk tolerance is emotional: how much volatility you can stomach. Risk capacity is structural: how much volatility your life can absorb without forcing bad decisions.

A household can have high risk tolerance ("we're fine with ups and downs") and low risk capacity (high fixed costs, low cash, employer-stock concentration). That combination is dangerous because it encourages you to take risks you can't actually afford to hold through.

Improving risk capacity is often less about changing investments and more about changing the household balance sheet: lowering fixed costs, increasing cash buffers, and reducing correlated risks.

To make stress-testing practical, treat it like an annual home inspection. You're not expecting the roof to collapse. You're checking for termites.

Here's a household-friendly stress-test checklist you can complete in an hour:

1. *Concentration:* What percent of investable assets is in any single stock (especially employer stock)? What percent is in one sector?
2. *Liquidity:* How many months of essential expenses can we cover without selling long-term assets?
3. *Fixed-cost pressure:* If one income drops for six months, which bills become immediately threatening?

4. *Tax friction:* If we needed $20,000 in cash this year, where would it come from, and what taxes might that trigger?
5. *Rebalancing reality:* In a 30% market drop, do we have a written, realistic rule for rebalancing—and will we follow it?
6. *Contribution resilience:* If contributions fall by 50% for a year, what is our fallback plan?

Notice what's not on the list: "Guess what the market will do." Stress-testing is about preparation, not prophecy.

Actionable takeaway: schedule a 60-minute "portfolio stress test" this month and run one brutal scenario: a 35% stock decline *plus* one income loss lasting six months. Write down (1) your employer-stock percentage, (2) your essential monthly expenses, (3) your liquid cash runway, and (4) exactly which account you would tap first if cash ran short. If any single stock exceeds your comfort cap or if your cash runway is under your target, make one concrete change within the next two pay cycles-either reduce the concentration, raise the cash buffer, or lower a fixed cost-so the fortress gets stronger in a way you can feel.

7.6 Automating Your Escape Plan

The most dangerous moment in household finance isn't a market crash. It's a random Thursday when you're exhausted, the kids are melting down, a work deadline is screaming, and you realize you haven't moved money into savings in three months.

Not because you're irresponsible—because you're human, and life is loud.

Automation is how you make your best intentions louder than your calendar.

It's tempting to treat "automating finances" as a techy bonus feature, like getting a smart thermostat. But for a two-income household trying to avoid the trap, automation is structural. It's a way to keep progress going even when your attention is hijacked by school forms, car repairs, and the endless churn of adult life.

A good system doesn't require motivation. It requires *defaults*.

Defaults are powerful for the same reason gravity is powerful: you don't have to believe in them for them to work. If your paycheck lands and nothing happens next, spending expands to fill the space. If your paycheck lands and money automatically moves to the right places, your lifestyle adapts around what remains.

That's not deprivation. That's design.

The best households don't have more willpower than everyone else. They have fewer moments where willpower is required.

Automation is also a marriage strategy. Many couples don't fight about money because they disagree on the big picture. They fight because money requires constant micro-decisions: "Did you move the transfer?" "Should we invest now or wait?" "Are we behind?" "Why is the checking account low?" Micro-decisions create micro-friction, and micro-friction accumulates.

When the system is automatic, the default outcome is "we're on track," and your money conversations can shift from firefighting to occasional steering.

To automate well, think in one simple flow: *income → bills → buffers → investing → guilt-free spending.*

Most households accidentally run the opposite flow: income → spending → leftovers (maybe) → investing (if we remember). That's a system built on hope. Hope is not bankable.

The first automation lever is the least glamorous and the most impactful: *separating bill money from life money.*

If all paychecks land in one checking account and everything is paid from there, the household is forced to constantly interpret a single number: "Is this enough?" That question is a stress factory.

A calmer setup uses at least two checking-style buckets:

- a *Bills account*: mortgage/rent, utilities, childcare, insurance, minimum debt payments.
- an *Operating account*: groceries, gas, family life, fun, the normal mess.

Your employer direct deposit can do some of this automatically. If not, an automatic transfer can.

The goal is psychological as much as mathematical: you want to stop having the "can we afford it?" conversation every time you buy paper towels. Bills get paid because bills money is protected. Life spending is constrained because it has its own lane.

Next, automate the buffers: emergency fund and near-term sinking funds.

A sinking fund is just savings with a job: car replacement, home repairs, holiday travel, medical out-of-pocket costs, the next inevitable "surprise." Without sinking funds, every surprise becomes a credit card event or a fight. With sinking funds, surprises become mildly annoying calendar items.

Automation makes sinking funds work because it turns "we should save for that" into "we are saving for that," without requiring a monthly debate.

A practical rule is to set these transfers for the day after payday. Not the end of the month. Not "when we see what's left." Money left in checking is money that will volunteer for a new purpose.

Then automate investing. The simplest and most effective version is: contribute to workplace retirement accounts automatically, invest in a taxable brokerage automatically if you need additional goals, and set a schedule for any annual increases.

This is where two-income households can create a quiet superpower: *auto-escalation*. Auto-escalation means your contribution rate increases gradually over time, often timed to raises. Many people intend to "invest more later," but later is busy and expensive. Auto-escalation makes later automatic.

A useful household approach is to tie increases to events that already happen:

- every annual raise: increase retirement contributions by 1–2 percentage points

- every debt payoff: redirect the freed payment into investing
- every childcare milestone: when daycare ends, redirect some of that cash flow before lifestyle absorbs it

That last one matters because childcare is one of the largest temporary expenses families face. When it disappears, it can feel like you got a "raise." If you don't claim that "raise" deliberately, it will vanish into better groceries, nicer vacations, and a slightly more expensive daily life that becomes your new baseline.

Automation lets you capture it before it evaporates.

A real-world example shows what this looks like in a household that isn't trying to be perfect.

Andre and Kim (composite) both work demanding jobs and have two school-age kids. They used to "do their finances" on Sundays. That lasted three weeks. After that, money management happened whenever someone felt anxious, which meant it happened at the worst times.

They built a system that took one evening to set up and about ten minutes a month to maintain:

- Paychecks: both direct deposits split automatically—most into Bills, the rest into Operating.
- Bills: all fixed bills on autopay from Bills, timed right after payday.
- Buffers: $300 per paycheck into an emergency fund until it hit their target; then that same transfer rerouted to home repairs.

- Investing: 401(k) contributions set and forgotten; Roth IRA contributions automated monthly; a small taxable auto-invest for medium-term flexibility.
- Spending: a weekly "allowance" transfer into Operating, so the week had a boundary.

The surprising result wasn't just that they saved more. It was that money stopped being a background hum of stress. They still had expenses and surprises, but the system absorbed them with less drama. Their marriage didn't become a finance podcast. It became quieter.

Now, what about robo-advisors?

A robo-advisor is a platform that builds and manages a diversified portfolio for you, typically using low-cost funds, automatic rebalancing, and sometimes tax features like tax-loss harvesting (selling investments at a loss to offset gains, within tax rules). Robo-advisors can be a great fit for households that want to outsource portfolio maintenance and reduce decision fatigue.

The benefit is not that a robo-advisor is magically smarter than you. The benefit is that it can be *consistently boring* in a way humans struggle with. It rebalances without ego. It doesn't read headlines. It doesn't panic sell because a coworker panicked.

That said, automation is only as good as what you automate. If you automate a bad default, you lock in a bad outcome more efficiently.

So if you use a robo-advisor (or any managed solution), the household constitution still matters. You want to verify a few key things:

- *Costs:* the advisory fee plus the underlying fund fees.
- *Asset allocation:* does it match your household risk capacity?
- *Tax approach:* does it coordinate with your other accounts, or does it create complexity?
- *Behavioral fit:* does the platform reduce your urge to tinker, or encourage it with constant performance updates?

The best platform is the one that helps you stick to your plan with the least friction.

Target-date funds deserve mention here because they are the default automation inside many 401(k)s.

A target-date fund is a one-fund portfolio that automatically shifts (via a "glide path") toward more conservative holdings as the target year approaches. For households who want simplicity, a target-date fund can be an elegant solution: diversification, rebalancing, and gradual risk adjustment in a single holding.

But target-date funds are not all the same. Two funds with the same year on the label can have very different stock/bond mixes and very different fees. Some are designed to reach a conservative allocation *at* retirement ("to" retirement), while others continue adjusting *through* retirement ("through" retirement). The labels look similar; the guts can be different.

The automation lesson is simple: don't just accept the default because it's the default. Accept it because you checked that it fits.

Automation should also include "avoid dumb mistakes" workflows, because households lose money not only from markets but from admin errors.

Examples:

- Set beneficiary reviews on a calendar (life changes; old beneficiaries create heartbreak and legal chaos).
- Keep a simple account inventory (where accounts are held, logins, two-factor authentication recovery, contact numbers).
- Use strong password management (financial identity theft is a brutal distraction).
- Turn on notifications for large transactions and transfers.

This is not paranoia. It's basic hygiene—like locking your doors. When you automate hygiene, you reduce the odds that a busy season becomes an expensive season.

There's also a subtle automation win that high-income two-income households often overlook: automating *fee minimization.*

Fees are not only about fund expense ratios. They include advisory fees, account fees, trading costs, and the "activity tax" of constantly making changes. Automation can reduce fees by keeping you invested in low-cost options and reducing the temptation to chase complex products.

A household that sets up automatic investing into a low-cost diversified fund is not just investing. It's quietly refusing a whole industry's worth of expensive distractions.

One caution: automation should not make you blind.

You still need a rhythm of review. Think of it like autopilot on a plane: the pilot doesn't stop paying attention; the pilot stops manually adjusting every second.

A practical review cadence for busy households:

- monthly (10 minutes): confirm bills cleared, transfers happened, and contributions are flowing
- quarterly (30 minutes): check allocation drift and rebalance only if your rules require it
- annually (60–90 minutes): update goals, adjust savings rate, review insurance, review beneficiaries, revisit the constitution for any major life changes

This keeps you in control without making you a full-time portfolio manager.

The final, most important automation is psychological: automate the decision to *keep going* when fear spikes.

If you can set contributions to happen without your mood's permission, you're building wealth in the background while your brain does whatever it does. That is the whole point of the escape plan: progress that doesn't require constant heroics.

Actionable takeaway: build your "payday cascade" in the next seven days. On the day after payday, automate (1) a transfer to your Bills account (if you split accounts), (2) a transfer to your emergency fund or sinking funds, and (3) an automatic investment into your chosen accounts. Then add one auto-escalation rule—*increase retirement contributions by 1% on the date of your next raise*—so your future self doesn't have to negotiate with your lifestyle when the money gets bigger.

Chapter 8

Financial Independence and the Long Game

You have learned how to earn, budget, and save together. Now comes the question that keeps even the most diligent savers awake at night: 'Is it enough?' This chapter shifts your mindset from accumulation to preservation and distribution. We will move beyond the vague goal of 'retiring someday' to building a concrete, two-income escape plan. You will learn how to turn your pile of savings into a paycheck that never quits, how to weather stock market storms without panic, and how to shield your hard-won nest egg from digital threats. This is not just about being rich; it is about being free.

8.1 The Alphabet Soup of Freedom: 401(k)s, IRAs, and HSAs

At some point, nearly every two-income household has the same quiet, ridiculous moment: you're doing "responsible adult" things—setting up payroll, picking benefits, clicking through an HR portal—and you realize you're about to make a decades-long decision based on a drop-down menu and a vague memory of something you once heard at a party about "Roth being better."

That's not a personal failure. It's a design feature of modern money. Retirement accounts are powerful precisely because they're wrapped in acronyms, rules, and fine print. But the goal is simple: keep more of what you earn, get free money when it's offered, and put your savings into containers that grow with fewer taxes and fewer leaks.

There are three main "containers" most households bump into: the workplace retirement plan (often a 401(k)), the individual retirement account (IRA), and the health savings account (HSA). Each one gives you a different kind of advantage—like choosing between three routes to the same destination. The trick is not memorizing every turn. The trick is knowing which on-ramp to take first.

The only guaranteed 100% return you will ever see

A 401(k) match isn't an investment strategy. It's a pay raise that only arrives if you claim it.

Here's how it works in plain language: you put some money into your 401(k) from your paycheck, and your employer adds extra money—often a percentage of what you contribute—up to a limit. If you contribute enough to reach the match threshold, you've effectively doubled (or partially doubled) that slice of savings instantly. The market can crash. Your coworker can brag about "timing the dip." None of that beats an immediate, contractual return on day one.

And yet, people miss it all the time—not because they're reckless, but because life is loud. Maybe cash flow is tight. Maybe the form was confusing. Maybe the match vests over a few years and the job feels temporary, so it doesn't feel "real."

A useful way to see it: if your employer matches 50% of the first 6% you contribute, and you earn $100,000, then contributing 6% ($6,000) gets you an extra $3,000. If you don't contribute, you don't just lose the "investment return." You lose compensation you already earned but failed to pick up.

Yes, vesting can complicate it. Some matches are yours immediately; others become yours gradually. But even with vesting, the match is usually the best first move, because you're starting with an advantage that investing alone rarely replicates.

And there's a second reason to grab the match early: it forces a healthy habit. Payroll deductions happen before your money has a chance to become "available" for impulse spending. You're not relying on willpower at the end of a long month. You're building a system.

Now for the part people don't say out loud: the match is also psychological protection. When markets drop, it's easier to keep contributing when you can tell yourself, truthfully, "I'm still getting paid to do this."

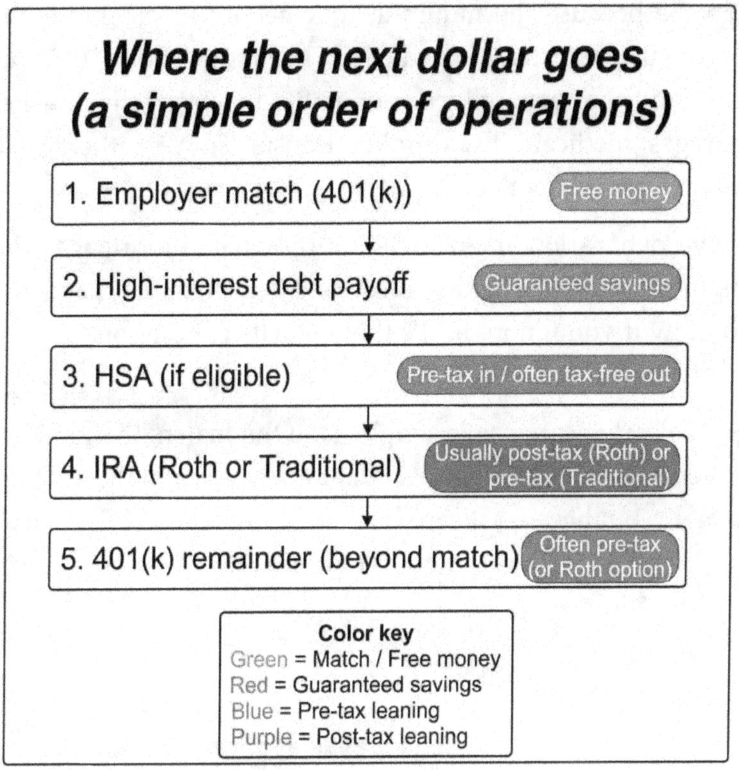

Before you invest harder, stop the financial bleeding

After the match, the next "investment" is often not an investment at all: it's paying off high-interest debt.

If you're carrying a credit card balance at 22% APR, paying it off is like earning a risk-free 22% return. You don't get a ticker symbol. You don't get bragging rights at a barbecue. You get something more valuable: the monthly payment disappears, and your future paychecks get room to breathe.

This matters more in a two-income household than people admit. Two incomes can create the illusion that debt is manageable because the minimum payments fit. But minimum payments are a stealth fixed cost. They reduce your ability to save, and they also reduce your ability to absorb shocks—job changes, medical bills, a broken transmission—without sliding into worse debt.

If the debt is low-interest (like a reasonable mortgage) the math changes. But high-interest debt is the one bill that tends to grow if you ignore it. It's the opposite of compounding in your favor.

So the early sequence is simple: take the match, then kill the expensive debt. Not because debt is "bad," morally or emotionally, but because it's expensive. The household that clears high-interest debt is the household that gets to keep more of its own future.

HSA: the account that behaves like a retirement cheat code

An HSA is weird in the best way. It sits in the benefits menu next to dental coverage and life insurance, so it feels like a minor HR detail. But financially, it can be one of the strongest long-game accounts available.

The basic idea: if you're enrolled in a qualifying high-deductible health plan (HDHP), you can contribute to an HSA. The money goes in with tax advantages, it can grow invested, and you can withdraw it for qualified medical expenses. In many cases, it's the closest thing most households get to a "triple tax advantage" setup: tax relief on the way in, tax-free growth while invested, and tax-free withdrawals for medical expenses.

Here's what makes it powerful for a two-income household: healthcare costs tend to show up no matter how smart you are. You can be a perfect saver and still have braces, a pregnancy, an injury, therapy, prescriptions, or just the slow creep of costs as you age. An HSA is a way to fund that reality with fewer taxes.

There are two common ways people use HSAs:

The spender approach. You contribute, then use the HSA debit card to pay current medical bills.

The investor approach. You pay current medical bills out of pocket (from your checking account), save the receipts, and let the HSA money stay invested for years. Later—sometimes much later—you reimburse yourself from the HSA for those past medical expenses. This turns the HSA into a flexible, long-term pool that can support healthcare spending in retirement.

That investor approach isn't for everyone. It requires cash flow and organization. But it's a real lever for households that can swing it: you're effectively letting money stay sheltered and compounding while you keep the receipts like IOUs to your future self.

One important nuance: you only get an HSA if your health plan qualifies. Not all plans do. If you have access, it often belongs near the top of your savings order because it's both tax-efficient and aligned with a cost category that tends to be unavoidable.

IRA: where you buy flexibility, not just tax benefits

A workplace plan is like a cafeteria: the options are chosen for you. An IRA is like a kitchen: you choose what goes in.

That freedom matters. Many 401(k)s are excellent, but some have limited fund options or higher fees. An IRA can widen your investment menu and lower your costs. It can also give you more control over whether you pay taxes now or later.

Two main flavors exist:

Traditional IRA. Contributions may be tax-deductible (depending on income and whether you have a workplace plan). You generally pay taxes when you withdraw in retirement.

Roth IRA. Contributions are made with after-tax dollars. Qualified withdrawals in retirement can be tax-free.

People get stuck trying to "solve" Roth vs. Traditional as if there's one right answer. Most of the time, the right answer is more like: *choose the one that fits your likely tax situation, then keep saving consistently.*

A helpful mental model is to treat it like choosing when to pay the tax bill:

Roth is paying taxes now to buy tax-free later. This can be attractive if you're early in your career, temporarily in a lower tax bracket, or you value the simplicity of tax-free withdrawals later.

Traditional is taking the tax break now and paying later. This can be attractive if you're currently in a high tax bracket and expect your taxable income to be lower in retirement.

But a two-income household has a special complication: your tax bracket can swing. Maybe both of you work now, but later one of you scales back, starts a business, or takes time off. Maybe one income grows quickly. Maybe a move changes state taxes. Your "future bracket" isn't a single prediction—it's a range of possible lives.

That uncertainty is a great reason to diversify *tax treatment*, not just investments. Many households end up with a mix: some money in pre-tax accounts (Traditional 401(k)/Traditional IRA), some in post-tax accounts (Roth), and potentially an HSA as a healthcare-specific bucket. That mix gives you options later—options are what freedom actually feels like.

A real-world case study: the couple who finally stopped ``saving last"

Consider Maya and Jordan, both 34, with two kids. Maya is a nurse, Jordan works in IT. Their combined income is solid, but their money always felt like it was evaporating. They were doing what many responsible people do: paying bills, trying not to overdraft, and "saving what's left." Which meant saving almost nothing.

They had these pieces on the board:

- A 401(k) at Jordan's job with a 4% match.
- Maya had a 403(b) (similar idea to a 401(k)) but no match.
- A credit card balance at 19% APR that had become a "permanent" monthly payment.
- An HDHP option with an HSA, but they'd ignored it because the deductible sounded scary.

Their turning point was not a complicated spreadsheet. It was a sequence.

First, Jordan set his 401(k) contribution to hit the full match. That was the easy win: immediate value, automatic, done.

Second, they redirected Maya's would-be retirement contribution temporarily toward the credit card payoff. This felt backward at first—like they were "falling behind" on retirement—but within eight months the card was gone. Their monthly cash flow improved permanently because the minimum payment vanished.

Third, they switched to the HDHP and started contributing to the HSA—modestly at first. The key move was psychological: they set a small separate "medical buffer" in their savings account so the deductible didn't feel like a cliff. With that buffer and the HSA contribution, the plan stopped feeling scary and started feeling structured.

Fourth, they opened a Roth IRA for Maya. Why Maya? Not because she was "behind," but because the IRA gave them flexibility and low-cost investment choices. They set up automatic contributions on payday, small enough to be sustainable.

Finally, once the debt was gone and the systems were in place, they increased Jordan's 401(k) contribution beyond the match. At that point, it didn't feel like deprivation. It felt like turning up the volume on a machine that was already running.

The headline isn't "they optimized taxes." The headline is: *they stopped asking willpower to do a system's job.* Each move reduced friction and increased momentum. That's what makes a plan durable.

401(k) beyond the match: still good, but watch the hidden drain

Once you've captured the match and handled high-interest debt, increasing your 401(k) contributions is often the next big engine—especially for higher earners who need a larger tax-advantaged lane.

But this is where you earn your keep as a financial adult: you look for the quiet leaks.

The biggest leak is fees.

Fees are not just a small annoyance. They are a permanent headwind. A difference of even half a percent in investment expenses, compounded over decades, can become a painful gap. Many employer plans are excellent and low-cost; some are not. The point isn't to panic—it's to be awake.

A practical way to assess it without becoming a hobbyist:

- Find the expense ratios of the funds you're using (often listed right in the plan).
- Prefer broad, diversified index funds when available (simple, low-cost, hard to game).
- Avoid "cash" or stable-value defaults unless you're intentionally choosing them.

Another leak is poor contribution pacing. Some people front-load contributions early in the year and accidentally miss match dollars if their employer matches per paycheck rather than "trueing up" at year-end. If your plan doesn't true up, maxing out too early can reduce the match you receive. This is one of those small-print details that can cost real money.

And then there's the human leak: panic.

A 401(k) is designed to be boring. The more dramatic it feels, the more likely you're checking it too often or taking market movement personally. If you're contributing steadily over decades, volatility is the price of admission for growth. The match is your friendly reminder that the system is paying you to stay seated.

Roth vs. Traditional: a way to decide without fortune-telling

Households often stall here because the decision feels permanent. It isn't. You can change future contributions as your situation changes. You can also end up with both types over a lifetime, which is often ideal.

If you want a clean, usable rule of thumb:

If money is tight and you need the tax break to keep saving, Traditional contributions can help you stay in the game.

If you can afford the taxes now and you like the idea of building a tax-free pile for later flexibility, Roth can be compelling.

And if you're still torn, consider splitting. Complexity isn't always good, but tax diversification is one of the few "hedges" ordinary households can build without getting fancy.

Also, remember the emotional reality of two incomes: one income can disappear faster than people want to believe (layoffs, burnout, caregiving). Building a mix of account types can help you adapt if your taxable income changes sharply in either direction.

The simplest order of operations you can actually follow

Financial advice fails when it's technically correct but too fragile for real life. The point of an "order" isn't perfection; it's avoiding the big regrets.

A household-friendly sequence looks like this:

1. Contribute enough to capture every dollar of employer match.
2. Pay off high-interest debt so your cash flow stops feeding interest.
3. If eligible, fund an HSA (especially if you can invest the balance).
4. Fund an IRA for flexibility and typically broader, lower-cost choices.
5. Increase 401(k) contributions beyond the match for higher-volume, tax-advantaged saving.

If you do those steps imperfectly, you still win. The magic isn't in picking the perfect fund. It's in consistently moving dollars into the right containers, in the right order, before life spends them for you.

Actionable takeaway: one hour, one paycheck, one win

Pick one paycheck in the next two weeks and do a single, concrete action that permanently improves your trajectory:

- If you're not getting the full 401(k) match: increase your contribution to the match point today.

- If you are getting the match: choose one high-interest debt and set an automatic extra payment that will pay it off measurably faster.
- If you have an HDHP option and haven't evaluated the HSA: compare plans and decide whether an HSA contribution—even a small one—fits your cash flow.

The goal is not to become an acronym expert. The goal is to make sure the next dollar you save goes to work immediately—either earning free money, avoiding expensive interest, or growing in the most tax-friendly container you can access.

8.2 How Much Is Enough? Finding Your Number

The most expensive sentence in personal finance is: "I think we're doing okay."

It sounds calm. Mature, even. It's also how perfectly competent, hardworking couples drift for a decade—saving a little when they remember, investing when the market feels friendly, and never quite knowing whether they're building a bridge to freedom or just stacking bricks in the backyard.

The internet loves a clean number. "You need $2 million." "You need 25 times your salary." "You need to max everything." Those slogans work the way horoscope readings work: sometimes they feel right, and they're almost always vague enough to avoid being falsified.

A real number—the one that changes your behavior in a useful way—comes from your household spending. Not your income.

Not your title. Not the vibes of your friend who moved to Portugal. Spending is the fuel you burn to live your current life, and your "enough" is simply the size of the tank you need to keep buying that fuel after paychecks stop.

That's the entire game. The rest is math, honesty, and a little humility about the future.

The quiet shift: from a savings goal to a spending plan

Most people start with a savings rate because it feels like virtue. "We save 10%." "We're trying to get to 15%." But a savings rate without a destination can make you do strange things. You can save 20% and still be financially fragile if your fixed costs are huge and your lifestyle quietly inflates every year. Or you can save 8% and be surprisingly resilient if your spending is low, your systems are solid, and your future obligations are predictable.

A spending-based number does something emotionally important: it turns retirement from a foggy concept into a solvable problem. You don't need to know what the market will do next year. You need to know what your household actually costs to run, and which parts of that cost are non-negotiable.

There's also a relationship dynamic buried in here. Two-income households often carry a hidden tension: one partner thinks "we're fine," the other feels like the floor could drop out at any moment. A shared number doesn't eliminate anxiety, but it gives you a common language. You're no longer arguing about feelings. You're comparing assumptions.

Step one: estimate your retirement spending (and admit what changes)

The simplest approach is to start with what you spend now. Not what you *wish* you spent. Not the month you had no surprises. A real average.

If your household uses a budgeting app, you're already holding the raw materials. If you don't, bank and credit card statements still tell the story with brutal clarity. Pull the last 12 months, categorize the big rocks, and compute an average monthly spend.

A practical scaffold is to think in the same buckets most households already live in:

- *Housing:* mortgage/rent, property taxes, insurance, repairs, HOA.
- *Transportation:* car payments, gas, maintenance, insurance, transit.
- *Food:* groceries, restaurants, coffee runs that pretend to be "just coffee."
- *Healthcare:* premiums, out-of-pocket costs, prescriptions, dental, vision.
- *Insurance and protection:* life, disability, umbrella, long-term care (if you choose it).
- *Kids and family:* childcare, activities, tuition help, travel to see relatives.
- *Fun and meaning:* travel, hobbies, giving, the things that make life feel like yours.

Then you do the adult part: you decide what changes in retirement.

Some costs often drop:

- Payroll taxes stop on portfolio withdrawals.
- Work costs often shrink (commuting, wardrobe, convenience spending).
- Saving for retirement stops once you are, in fact, retired.

Some costs often rise:

- Healthcare can become a larger line item, especially before Medicare eligibility.
- Travel and leisure can increase if you actually intend to enjoy your time.
- Home maintenance doesn't retire when you do.

The point isn't to predict perfectly. The point is to stop using mystery as a strategy.

A useful way to keep yourself honest is to build two numbers:

Baseline spending: the "keep the lights on" version of your life.

Comfortable spending: baseline plus the things you genuinely want (travel, dining out, hobbies, helping family) without constant negotiation.

That second number matters because many couples accidentally plan for baseline and then feel guilty for living like humans. Planning for joy is not indulgent; it's realism.

The 4% rule: what it is, what it isn't, and why it's still useful

You'll hear the "4% rule" described as if it's a law of physics. It isn't. It's a rule of thumb—a starting point—that came from studying historical market returns and asking a simple question: if a retiree withdraws a fixed percentage of their starting portfolio each year (adjusting for inflation), how often does the money last for a typical retirement length?

The reason it became famous is not because it's perfect, but because it's memorable and surprisingly practical. It gives you a quick translation:

$$\textit{Financial Independence Number} \approx \frac{\textit{Annual Spending}}{0.04} = 25 \times \textit{Annual Spending}$$

If your household expects to spend $80,000 per year, the back-of-the-envelope number is:

$$\$80{,}000 \times 25 = \$2{,}000{,}000$$

That's the headline. The more interesting part is what's hidden underneath.

The 4% rule is not a promise. It doesn't guarantee you'll never run out. It doesn't guarantee you'll be comfortable. And it doesn't mean you should withdraw exactly 4% forever no matter what happens.

What it *does* offer is a disciplined way to stop guessing. It ties the size of your portfolio to the lifestyle it needs to support.

And it points to the real risk retirees face: not whether markets go down (they will), but whether you retire into bad market years early on and keep spending as if nothing happened. That "sequence" problem—the order of returns—matters more than the average return. Your number is not just a destination; it's a relationship between your spending and your flexibility.

So treat 4% like a compass, not a calendar. It tells you direction.

Why two-income households have a unique edge (and a unique blind spot)

Two incomes create a built-in buffer during the saving years. When one income wobbles, the other can keep the machine running. That's a real advantage.

But there's a matching blind spot: lifestyle expansion becomes easier to justify. A nicer car feels affordable because it's "only" one of the paychecks. Private school feels manageable because the second income covers it. A bigger house looks fine because the payment still fits under the combined total.

That's the two-income trap in its most polite form: you don't feel reckless, you feel normal.

Finding your number exposes this gently but firmly. If the lifestyle you built requires $160,000 per year to maintain, then the portfolio required to support that lifestyle is correspondingly huge. There's no shame in that. But there *is* power in seeing it clearly—because the easiest way to "earn" a million dollars isn't necessarily investing harder. Sometimes it's deciding not to spend the $40,000 a year that demands an extra million in the first place.

A case study: the couple who discovered they were aiming at the wrong target

Talia and Evan are both 41. Two kids, suburban life, good jobs. They assumed they needed "around two million" because that's what everyone says. They were saving consistently, but the number felt abstract—like a mountain in the distance that never got closer.

When they finally calculated their spending, the first surprise was how high it was: $11,500 per month, or about $138,000 per year. Not because they were extravagant, but because their fixed costs had quietly become enormous: a large mortgage, two car payments, high insurance premiums, and childcare that cost more than a small rent payment.

Their second surprise was emotional: they didn't even *like* some of the things that made their spending high. They liked their neighborhood, yes. They didn't like the stress of juggling two car loans, nor the constant feeling that every month required perfect execution.

Now the math did what math does—it told the truth without drama:

$$\$138{,}000 \times 25 = \$3{,}450{,}000$$

That wasn't their "failure." It was their starting point.

Then they built their baseline and comfortable budgets.

Baseline (keep life stable): $105,000/year.

Comfortable (travel, extras, helping family): $125,000/year.

Those two numbers reframed everything. Their "enough" wasn't a single mountain; it was a range:

$$\$105{,}000 \times 25 = \$2{,}625{,}000$$

$$\$125{,}000 \times 25 = \$3{,}125{,}000$$

Next came the lever that actually moved their future: fixed costs.

They refinanced nothing (rates weren't favorable). They didn't do a dramatic downsizing. They did something less cinematic and more effective: when one car died, they replaced it with a reliable used vehicle and kept the other until it was truly done. They redirected the freed-up payment into investments automatically. They also committed to a "household baseline" rule: any new recurring cost had to come with an explicit plan for what it replaced.

Their spending dropped by about $1,200 per month without touching the things they cared about most. That's $14,400 per year.

On paper, that sounds like a budgeting win. In reality, it was a number win:

$$\$14{,}400 \times 25 = \$360{,}000$$

They didn't just save $14,400. They reduced the portfolio they needed by roughly $360,000.

That's what makes spending-based planning feel almost unfair—in the best way. It rewards clarity.

Make your number more realistic by subtracting future income floors

A common mistake is to calculate the full spending need and assume the portfolio must cover every dollar forever. Many households will have other income sources later—most notably Social Security.

A simple framework:

$$\textit{Portfolio Spending Need} = \textit{Annual Spending} - \textit{Reliable Annual Income Floors}$$

If you estimate \$90,000/year in retirement spending and expect \$35,000/year from Social Security (combined, inflation-adjusted estimate), then your portfolio needs to fund:

$$\$90{,}000 - \$35{,}000 = \$55{,}000$$

Using the same rule of thumb:

$$\$55{,}000 \times 25 = \$1{,}375{,}000$$

This is a dramatically different target than \$2.25 million (\$90,000 × 25). Not because Social Security is a "bonus," but because it's an income stream that can cover part of your baseline.

The nuance is timing. Social Security might start later than your retirement date, and claiming early vs. later changes the amount. So it's often wise to do two calculations:

Before Social Security begins: larger portfolio draw.

After Social Security begins: smaller portfolio draw.

That gap is where planning becomes practical instead of theoretical. You might decide you need a bigger cash cushion for early retirement years, or a part-time bridge, or simply a more conservative withdrawal early on.

Add a margin of safety without turning your life into a spreadsheet

People hear "retirement planning" and imagine precise forecasts, as if you can model 30 years of markets, inflation, health, taxes, and family obligations with a tidy formula. You can't.

But you can build margin like a grown-up.

Three places margin tends to matter most:

Healthcare: If you retire before Medicare age, premiums and out-of-pocket costs can be a shock. Build a specific healthcare line item instead of hoping it "works out."

Home and maintenance: Roofs, HVAC systems, and plumbing don't care about your asset allocation. If you own a home, include an annual maintenance reserve.

Family: Many two-income households end up supporting aging parents, adult children, or grandchildren in ways they never predicted at 30. You don't have to fund every possible scenario, but it's wise to decide what help you want to be able to offer.

Margin doesn't mean hoarding. It means you don't have to panic when life behaves like life.

A number you can use tomorrow morning

The best "finding your number" exercise isn't the one that creates a perfectly accurate target. It's the one that changes what you do next week.

Here's a simple version that's shockingly effective:

1. Calculate your household's average annual spending from the last 12 months.
2. Build two retirement spending estimates: baseline and comfortable.
3. Use 25× spending as a first-pass portfolio target.
4. If you expect Social Security or a pension, subtract a conservative estimate from spending and recalculate the portfolio target.

Then, one step that turns theory into traction: write down *one* spending decision that reduces your target by at least $100,000 without lowering your quality of life.

You do that by lowering recurring spending by $4,000 per year (because $\$4{,}000 \times 25 = \$100{,}000$). That might be a car choice, a housing choice, or a subscription ecosystem that's gotten out of hand. The point is to pick something that repeats every year—because recurring costs are what inflate your "enough."

Actionable takeaway: calculate your number in 30 minutes, then make one permanent change

Set a timer for 30 minutes. Pull your last three months of bank and credit card statements and write down your monthly average spending in five lines: housing, transportation, food, healthcare, everything else. Multiply by 12 for an annual estimate.

Then do two quick calculations on paper:

$$\textit{Baseline Number} = 25 \times (\textit{Annual Spending} \times 0.85)$$

$$\textit{Comfortable Number} = 25 \times (\textit{Annual Spending} \times 1.00)$$

Those percentages aren't sacred—they're a way to force two scenarios without getting stuck. Once you see the range, choose one recurring expense to redesign in the next 30 days, specifically because it shrinks the size of the portfolio you need to buy your freedom.

8.3 The Safety Net: Social Security as Your Floor

A funny thing happens when people talk about Social Security at a dinner table. Someone will say, "I'm not counting on it," with the same tone you'd use for a raffle ticket. Everyone nods, as if cynicism is a financial plan.

But Social Security isn't a raffle. It's closer to a pension you earn in installments—one that follows you job to job, one that keeps paying as long as you're alive, and one that is deliberately designed to be more protective for lower and middle earners than most people realize. If you're building financial independence, the smartest way to think about Social Security is not as "extra." It's your floor: the cash flow that keeps the lights on so your investments don't have to carry the entire weight of your life.

That shift—*from bonus to foundation*—changes how you save, how you invest, and how much risk you actually need to take.

The floor-and-ceiling problem every household faces

Every household has two kinds of expenses.

There are *floor expenses*: housing, utilities, food, insurance, basic transportation, basic healthcare. The costs that show up even when your mood is terrible and your luck is worse.

And there are *ceiling expenses*: travel, dining out, gifts, hobbies, the nice version of everything, the "we can swing it" choices that make life feel spacious.

Financial independence gets dramatically easier when you stop trying to fund the floor with fragile money.

A portfolio is powerful, but it's moody. Markets drop. Inflation spikes. You might retire into a recession. If the portfolio is responsible for your survival spending, you feel every downturn in your nervous system. But if your survival spending is largely covered by guaranteed income—Social Security, maybe a pension—then the portfolio becomes what it was always meant to be: the flexible layer that funds comfort, joy, and optionality.

This is why Social Security deserves respect even from people who consider themselves "serious investors." It's not competing with your investing plan. It's stabilizing it.

What Social Security really is (in plain English)

Social Security is an inflation-adjusted lifetime benefit based primarily on your work history. The system looks at your earnings, applies a formula, and produces a monthly payment you can claim as early as age 62 or as late as age 70. If you're married (or were married long enough), there are spousal and survivor benefits that can matter just as much as your own work record.

Under the hood, it's complicated. On purpose. But the basic promise is simple:

- If you work and pay into the system, you earn future income.
- The benefit is designed to replace a higher share of income for lower earners than for high earners.
- The benefit continues as long as you live, which makes it a form of longevity insurance.

Longevity insurance is an unglamorous phrase that describes a terrifying risk: living a long time and running out of money. Social Security directly attacks that risk. Your portfolio can't guarantee a paycheck at age 92. Social Security can.

Why the benefit formula is more generous than most people assume

Social Security benefits are based on your *AIME* (Average Indexed Monthly Earnings). Translation: the system takes your earnings history, adjusts older earnings for wage growth (so a dollar earned decades ago isn't treated like a dollar earned last year), and then averages your highest 35 years of earnings.

Two details matter immediately for real households:

1. *Highest 35 years.* If you have fewer than 35 years of earnings, the missing years are treated as zeros in the average. This is a big deal for people who took time out of the workforce for caregiving, schooling, illness, immigration timing, or any of the normal chaos of adulthood.
2. *Indexing.* Earnings are generally wage-indexed through about age 60, which helps prevent early-career, low-salary years from permanently crushing your benefit.

Once you have AIME, the system uses a progressive formula to calculate your *PIA* (Primary Insurance Amount), which is essentially your monthly benefit at your "full retirement age" before early/late adjustments.

"Progressive" here means Social Security replaces a larger portion of income for lower earners. It's not flat. It's not "what you paid in, you get out." It's tilted toward making sure the floor exists.

To make that real without drowning in algebra, here's the intuition: the formula gives the most generous credit to your first layer of earnings, a moderate credit to the next layer, and a smaller credit above that. For people becoming eligible in 2026, those breakpoints (often called "bend points") are $1,286 and $7,749 of AIME, with different percentage factors applied to each layer. The exact math is less important than the shape: *the first dollars of lifetime earnings count more than the last dollars.*

That's why Social Security can be the financial glue for households with uneven earnings, career breaks, or a partner who did essential unpaid labor that made the other partner's career possible.

A case study: when one career ``counts" and the other one quietly disappears

Consider Denise and Mark, both 58. Mark worked continuously in a higher-paying field. Denise worked for years, then took a long break when their second child was born and later returned part-time. In their minds, Social Security was "Mark's thing," because Mark had the visible paychecks.

When they checked their records, Denise's estimated benefit on her own work history was modest—she had fewer than 35 strong earning years, so zeros and low-earnings years pulled down her average. She felt embarrassed, like she'd somehow failed a test.

Then they learned the part couples routinely miss: *spousal and survivor benefits can change the whole picture.*

Depending on their situation, Denise may be eligible for a spousal benefit based on Mark's record, and if Mark dies first, she may be eligible for a survivor benefit that replaces her own benefit. Suddenly, "Mark's Social Security" wasn't just his. It was a household asset with built-in protection for the surviving spouse.

This didn't magically solve retirement. But it did something just as important: it made their baseline safer. Denise stopped treating her future as a question mark, and Mark stopped treating the plan as "my account plus your account." They began thinking like a household with an income floor.

That's the emotional win: Social Security can reduce the fear that one partner's career path made the other partner permanently vulnerable.

Timing: why claiming age is a lever, not a trivia question

Social Security is one of the few retirement decisions where waiting can increase a guaranteed lifetime paycheck. Claim earlier, and your monthly benefit is lower. Claim later (up to age 70), and your monthly benefit is higher.

That doesn't mean "always wait." It means claiming is a tradeoff between:

- *Cash flow now* (useful if you retire earlier, have health issues, or need income to avoid draining investments), and
- *Bigger lifetime floor later* (useful if you expect a long life, want survivor protection, or want to reduce portfolio withdrawals in older age).

For two-income households, the claiming decision often becomes a coordination problem, not an individual one. Sometimes the higher earner delaying benefits can be a form of insurance for the surviving spouse, because the survivor benefit is tied to the deceased spouse's benefit amount. In plain language: *a higher earner waiting can protect the household's future "widow(er) income."*

This is where the "floor" idea becomes practical. If you can cover your baseline expenses from other sources for a few years—work, part-time work, cash reserves, a smaller portfolio draw—you might buy a permanently larger Social Security floor. If you can't, claiming earlier may be the right move. Either way, you're choosing, not drifting.

The uncomfortable question: ``Will it still be there?"

Social Security anxiety is not irrational. People hear the word "depletion" and picture benefits vanishing overnight.

The nuance matters: "trust fund depletion" does not mean "benefits drop to zero." It means that, under current projections and current law, the system's reserves could be exhausted around a certain year, after which ongoing payroll tax revenue would still cover a large portion of scheduled benefits—but not all of them—unless Congress changes taxes, benefits, or both.

To keep this grounded in actual dates: the 2024 Trustees Report summary projected depletion of the Old-Age and Survivors Insurance (OASI) trust fund in 2033. A Social Security Administration press release dated June 18, 2025 described the combined OASDI (retirement + disability) reserves as sufficient to pay scheduled benefits until about 2034, with roughly 81% of benefits payable thereafter from dedicated revenue, and noted the OASI depletion timing around 2033 with about 77% payable at that point under the projection.

You don't need to memorize those percentages. You need to use them to plan with adult caution instead of panic.

A practical planning approach is to treat Social Security like a floor, but apply a conservative haircut in your own math—especially for retirement plans that depend heavily on it. For example, if your estimate is $40,000 per year, you might plan on $30,000–$35,000 per year as the baseline floor in long-range projections. If reforms improve solvency, you've built upside. If benefits are reduced, you're not blindsided.

This is not pessimism. It's engineering.

How Social Security makes your investment plan sturdier

When households ignore Social Security, they often overburden the portfolio. They assume the portfolio must cover 100% of spending forever, so they chase a larger "number," take more market risk than they can stomach, or stay in a job they hate longer than necessary because the goalpost keeps moving.

When households integrate Social Security as an income floor, three good things happen:

1. *Lower withdrawal pressure.* The portfolio doesn't need to fund every dollar, which can reduce the withdrawal rate you require.
2. *More flexibility in bad markets.* If markets drop early in retirement, a stable floor means you can cut "ceiling" spending without threatening the basics.
3. *Better sleep.* This sounds fluffy until you live it. A guaranteed check changes the texture of retirement.

Think of it this way: the portfolio is a powerful engine, but Social Security is the suspension system. It makes the ride survivable when the road gets ugly.

Common mistakes that cost households real money

The most damaging Social Security mistakes are rarely about picking the "wrong" claiming age. They're about bad inputs.

Mistake 1: Never checking your earnings record. Errors happen. Missing years happen. If you wait until you're about to claim, you may be too late to fix problems easily.

Mistake 2: Underestimating the impact of zeros. If one spouse has a shorter work history, the 35-year averaging can reduce benefits more than people expect. Knowing this early can influence whether part-time work is worth it later, not just for current cash flow but for future benefits.

Mistake 3: Planning as two single people instead of one household. Spousal and survivor rules can materially change the safe floor for the surviving partner. Many couples accidentally optimize the wrong thing (like "who gets the biggest check first") instead of optimizing household security over a long life.

Mistake 4: Treating Social Security like gambling. Either "it'll be gone" (so you ignore it) or "it's guaranteed" (so you over-rely on it). Both extremes make planning weaker.

Turning Social Security into a clean household floor

Here's a way to make this concrete without getting lost:

1. Decide your *baseline annual spending* in retirement (the keep-the-lights-on number).
2. Estimate your combined Social Security benefit at a chosen claiming age (or a range).
3. Subtract a conservative amount of that estimate from baseline spending.
4. The remainder is what your portfolio must reliably cover.

Example: baseline spending is $72,000/year. You conservatively plan for $36,000/year from Social Security (after applying your own haircut). Your portfolio only needs to cover $36,000/year for the baseline.

That's not just "half the spending." It's half the stress. It's half the risk that a bad market year forces you to sell investments at the worst possible time.

Now your portfolio can fund the ceiling spending—the things you can adjust without endangering your basic dignity.

Actionable takeaway: build your floor this week

Do three things in one sitting—no spreadsheets required.

1. Create or log into your `mySocialSecurity` account and check that your earnings record looks accurate.
2. Write down two estimated benefit numbers for your household: one for a "claim earlier" scenario and one for a "claim later" scenario. Use the site's estimates as a starting point.

3. Pick a conservative floor number you're willing to plan around (even if that means taking 75%–85% of the estimate), and subtract it from your baseline annual spending.

The result is a single sentence you can actually use: *"Our portfolio must cover $X per year for the basics; Social Security covers the rest."* When you can say that sentence out loud—and both partners agree it's real—the future stops feeling like a rumor and starts feeling like something you can build.

8.4 The Art of the Withdrawal: Surviving Market Crashes

The cruelest retirement story isn't the one where someone never saved. It's the one where someone did everything "right," hit their number, finally quit, and then the market promptly threw a tantrum.

You can feel the scene: the first Monday without a commute, coffee tasting better, the strange joy of having a weekday to yourself. Then you check your account-because of course you do-and it's down 18%. Another month, down again. You haven't even bought the celebratory patio furniture and already it feels like you've made a mistake.

This isn't just nerves. There's a real mathematical trap hiding in the early years of retirement. It has an intimidating name—*Sequence of Returns Risk*—but it's basically one unfair rule: when you're withdrawing money, *bad returns early hurt more than bad returns later.*

Sequence of Returns Risk: the same average, wildly different outcomes

Most people understand investment risk as a long-run average: "Stocks return about X% over time." The problem is that retirees don't live in "over time." They live year to year, paying real bills with real dollars while the market does whatever it does.

Sequence of Returns Risk means that the *order* of market returns-when good years and bad years happen-can determine whether your money lasts, even if the *average* return over 30 years is identical.

If you get your worst market years at the beginning of retirement, you're selling shares while prices are depressed. That locks in losses and leaves you with fewer shares to recover when markets rebound. If you get those same bad years later, after a decade of growth, your portfolio has more cushion. Same average. Different life.

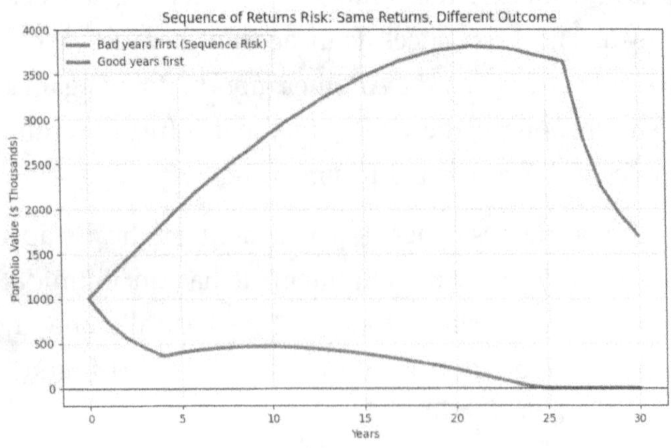

Once you see this, a lot of retirement advice gets reclassified. It's not that the famous "4% rule" is useless—it's that it assumes a level of steadiness that real life refuses to provide. Retirements don't fail because of one terrible year. They fail because of terrible timing plus rigid spending.

And that's good news, because timing is partly luck, but rigidity is a choice.

The dirty secret: retirees don't run out of money, they run out of flexibility

A traditional retirement plan is built around a heroic idea: pick a number, retire, and then spend the same inflation-adjusted amount every year as if the market is a calm employee who always shows up on time.

That plan works beautifully in spreadsheets. In the real world, it can turn you into the kind of person who is "retired" but afraid to replace a water heater.

A better approach is to treat retirement spending like driving in winter: you don't slam the brakes because the road is icy. You leave yourself room, you adjust speed, and you accept that the safest driver is the one who can react without drama.

That's where *guardrails* come in.

Guardrails: spending rules that behave like shock absorbers

Guardrails are a simple idea with a powerful emotional benefit: you pre-decide what you'll do if markets drop, so you don't have to negotiate with your fear later.

Instead of withdrawing a fixed amount forever, you set a spending range tied to your portfolio's health. If the portfolio falls below a certain threshold, you trim spending. If it rises above a certain threshold, you can loosen spending a bit (or bank the surplus). You're not timing the market. You're responding to it.

A household-friendly version looks like this:

- Pick a starting withdrawal (say, 4% of the portfolio).
- Decide what counts as "too low" (for example, if the withdrawal rate rises above 5% because the portfolio dropped).
- Decide what you will cut first, specifically.

The third bullet is where guardrails become real. "We'll cut spending" is not a plan. A plan is: "If markets are down and our withdrawal rate spikes, we pause big travel, reduce dining out, and delay the kitchen remodel." You're protecting the floor and trimming the ceiling.

If you did the work to define your baseline spending versus comfortable spending, you already have the blueprint. Guardrails are simply the agreement that *comfortable spending is allowed to flex.*

A case study: the couple who retired in the worst possible month (and stayed retired anyway)

Luis and Hannah retired in early 2008. Not because they were impulsive—because their plan said they could. They were 62 and 60, had a paid-off home, and had built a portfolio that looked, on paper, more than sufficient. They also had the normal retirement dream: travel, time with grandkids, a long-delayed sense of control over their days.

Then the market fell apart.

The first year, their portfolio dropped sharply while they were withdrawing for living expenses. Friends said, "Go back to work." They didn't want to, and realistically, Hannah couldn't-she was caring for her mom. Luis could have consulted, but the jobs available weren't the kind that would cover the stress.

What saved them wasn't brilliance. It was structure.

Before retiring, they'd divided their spending into two piles:

Baseline: property taxes, utilities, groceries, insurance, basic healthcare, and a modest "life is still fun" amount.

Extras: travel, gifts above a set amount, home projects, and the "why not?" spending that makes retirement feel celebratory.

They also kept two buffers:

A cash bucket: about a year of baseline expenses in cash-like holdings so they weren't forced to sell stocks at the bottom for groceries.

A flexible work option: not a full job plan, but a short list of ways Luis could earn some income if needed.

When markets dropped, they did three things quickly:

1. They cut extras, immediately and without resentment, because the rule was agreed on ahead of time.
2. They used the cash bucket to reduce the need to sell depressed investments.

3. Luis took small consulting work selectively, not because they were "failing," but because it let them leave investments alone.

It wasn't fun. But it worked. And here's the part that matters: when markets recovered, they didn't permanently ratchet their lifestyle down. They simply avoided doing long-term damage in the worst years.

Retirement didn't end. It got narrower for a while. That's the realistic version of success.

The three levers that matter most in a crash

When the market falls early in retirement, you have only a few levers that create outsized impact. None are glamorous. All are effective.

1) The size of your "must-pay" budget

The smaller your baseline, the easier it is to survive bad years. This is why households that reduced fixed costs on the way to retirement tend to feel calmer later. It's not virtue; it's physics. A low baseline reduces the amount you *must* withdraw.

If your baseline is $50,000 and your comfortable spending is $80,000, you have room to cut $30,000 in a bad year without threatening your core life. If baseline is $75,000 and comfortable is $80,000, you can't cut much at all. The market becomes your boss.

2) The cash and "boring money" buffer

A cash buffer-money that isn't exposed to stock market swings-can help you avoid selling stocks after they've fallen. This is not about hoarding cash forever. It's about giving your portfolio time to recover.

Even a modest buffer can change the psychology of retirement. Without it, every downturn feels like you're being forced to liquidate your future at a discount. With it, you have breathing room.

The right size depends on your household, your risk tolerance, and how flexible your spending is. Some households keep several months of baseline expenses; others keep a year or two. The goal is not to "beat the market." It's to make sure the grocery money doesn't depend on today's stock price.

3) The ability to earn *some* income

Many retirees hear "go back to work" as an insult, as if it invalidates retirement. But the ability to earn even $10,000–$20,000 in a bad market year can dramatically reduce the withdrawal burden on the portfolio.

It also changes the emotional story. You're not helpless. You're choosing to take pressure off your investments. For two-income households, this can be especially powerful if one partner enjoys part-time work or consulting more than the other. The point isn't equality. It's resilience.

Taxes and account rules: the invisible hand on your withdrawals

Withdrawal planning is not just "sell some investments." The account you withdraw from affects taxes, which affects how much you can safely spend.

A few realities that surprise people:

- *Pre-tax accounts* (like a traditional 401(k) or traditional IRA) generally create taxable income when you withdraw.
- *Roth accounts* can provide tax-free withdrawals if rules are met, which can be valuable flexibility in years when you want to control taxable income.
- *Required Minimum Distributions* (*RMDs*) can force withdrawals from certain pre-tax accounts starting at age 73, whether you need the money or not. Forced withdrawals can push you into higher tax brackets, increase Medicare-related premiums, and generally reduce your ability to "smooth" taxes.

This is one reason tax diversification matters. If all your money is in one tax bucket, your retirement becomes less steerable. A mix of pre-tax, Roth, and taxable accounts can let you choose the least painful source of cash in a given year.

Even if you never become a tax nerd, the concept is worth adopting: *the best withdrawal plan is the one that gives you choices.*

A practical guardrail system a household can actually follow

You don't need a complicated model to make this real. You need rules that can survive a stressful year.

Here is a simple, household-usable framework:

1. *Define the floor.* Write down your baseline annual spending (the "no drama" life).
2. *Pick a default withdrawal.* Start with a reasonable percentage of your portfolio for total spending in normal years.
3. *Set a cut trigger.* If your portfolio drops such that your withdrawal rate rises above a higher threshold, you automatically cut discretionary spending by a pre-set amount.
4. *Set a recovery rule.* You only restore discretionary spending after the portfolio recovers beyond a healthier threshold for a full year.

The recovery rule matters because without it, households whipsaw: cut spending in a bad year, then immediately inflate spending after a rebound, often before the portfolio has rebuilt a real cushion.

This is also where couples can avoid conflict. The fight in a downturn is rarely about money; it's about uncertainty and control. Guardrails convert uncertainty into a script. Not a perfect one, but one you agreed on when you weren't scared.

Actionable takeaway: write your "bad market year" script while you're calm

Within the next week, write a one-page retirement crash plan. Literally one page.

Include:

- Your baseline monthly spending number.
- A list of 5 discretionary expenses you will pause or shrink first in a market downturn (specific items, not vague categories).
- The size of the cash buffer you want to maintain (in months of baseline spending).
- One realistic "income valve" option (consulting, part-time work, seasonal work) that could bring in $10,000–$20,000 in a bad year without derailing your life.

Then put it somewhere you can find it, because the moment you need it is the moment you won't feel like inventing it. The goal isn't to predict the next crash. It's to make sure the next crash doesn't get to vote on whether you stay retired.

8.5 Locking the Vault: Digital Defense

A strange kind of theft has become normal: someone you've never met, in a place you'll never visit, quietly taking your money without ever touching your front door.

No broken window. No masked stranger. Just a text message that looks like your bank, a password you reused years ago, a "helpful" customer-service call that's actually a trap. And because it's invisible, it feels like something that happens to other people—until the day it happens to you, and then it feels personal in a way that's hard to describe. Not just anger. Violation. Embarrassment. The sickening thought: *I was careful. How did this happen?*

Here's the uncomfortable truth: if you're building wealth, you're becoming a better target. The same systems that make modern money convenient—instant transfers, app logins, digital signatures—also make it fast to steal. The goal of digital defense isn't to become paranoid. It's to make your household a *harder* target than the next one.

Cybersecurity can sound like a different universe, full of jargon and hoodies. But for a household, it boils down to three outcomes:

- prevent account takeover (someone logging in as you),
- prevent new credit fraud (someone borrowing as you),
- detect and contain damage quickly (so one breach doesn't become a financial wildfire).

If you do those three well, you can keep investing, saving, and living without the constant background fear that your money is one click away from disappearing.

The new weak link isn't technology. It's urgency.

Most successful scams don't "hack" your bank. They hack your attention.

They create urgency: *Your account is locked. Your package can't be delivered. Someone tried to log in. Your boss needs gift cards. Your payroll needs confirmation.* The message is designed to trigger the part of your brain that wants to fix problems immediately. And in a two-income household—two careers, kids, calendars, obligations—urgency works because everyone is busy.

That's why the most important security skill is not technical. It's behavioral: any message that pressures you to act fast deserves to be treated as suspicious, even if it uses your name and looks official.

A useful household rule is almost comically simple:

> If it's urgent, it's probably lying.

You can still respond. You just respond on *your* terms: by contacting the institution through a known number or bookmarked website, not by clicking the link or calling the number in the message.

Case study: the ``bank fraud call" that cleaned out a checking account

Renee and Sam had what most people would call a solid setup: direct deposit, automatic bill pay, investment contributions on schedule. They weren't reckless. They were efficient.

Sam got a call at 3:40 p.m. from "Fraud Prevention." The caller knew Sam's name, the last four digits of a card, and said there were suspicious charges. Sam was told they needed to "verify identity" and "secure the account." A text arrived with a one-time passcode. The caller asked Sam to read it aloud to confirm.

Sam did. Because it sounded like the responsible thing to do.

That code wasn't identity verification. It was the key to the account. The scammer used it to log in and add a new transfer destination. Within minutes, money moved out. Then came the second wave: the scammer called again to "help recover funds," guiding Sam through steps that actually authorized additional transfers.

When Renee came home, the money was already gone.

They did eventually recover some funds, but the damage wasn't just the dollars. It was the time: hours on the phone, accounts frozen, bill payments failing, payroll rerouted, credit monitoring, anxiety spikes every time a phone rang. The household's money system—something they'd built carefully—was suddenly fragile.

What would have stopped it? Not genius. Two simple defenses:

- never share one-time codes with anyone (even "your bank"),
- hang up and call back using the number on the back of the card.

That's the theme of digital defense: most losses come from a handful of predictable failure points. Patch those, and your risk drops sharply.

The password problem: your brain is not a vault

People reuse passwords for the same reason they eat dinner over the sink: it's faster. But password reuse is how breaches spread. If one site leaks your login, attackers try the same email and password everywhere—banking, shopping, even your email itself. Once your email is compromised, password resets become a highway into everything else.

So the goal is not "strong passwords." The goal is *unique passwords,* because uniqueness prevents a leak in one place from becoming a catastrophe everywhere.

That's why password managers are a household superpower. They let you:

- generate long, random passwords you'd never remember,
- use a different password for every account without losing your mind,
- share access safely between partners when appropriate (instead of texting passwords).

One household nuance matters: shared logins create messy risk. If both partners use the same credentials for financial accounts, you can't easily tell who did what, you're more likely to share codes verbally, and you're more likely to keep "temporary" access forever. Where possible, each partner should have their own login with their own authentication, especially for banks and brokerages. If a login must be shared, store it in the password manager and treat it like a hazardous material: no texts, no email, no sticky notes.

MFA: good, better, best (and why ``best" matters for money)

Multi-factor authentication (MFA) means you need more than a password to log in-usually something you *know* (password) plus something you *have* (a code or device). This is non-negotiable for financial accounts.

But not all MFA is equally strong.

Text-message codes (SMS) are better than nothing, but they can be vulnerable to SIM-swap attacks and other forms of interception. App-based authenticators are generally stronger because they're not tied to your phone number. Stronger still are security keys or passkeys (depending on the service), which are designed to be phishing-resistant—meaning a fake website can't easily trick you into giving away the login in a reusable way.

You don't need to become an expert in authentication standards to benefit from this. You just need a preference order:

- use a phishing-resistant method when offered (security key / passkey),
- otherwise use an authenticator app,
- use SMS only if it's the only option.

And one rule that belongs on the refrigerator:

> No legitimate institution will ask you to read them your one-time code.

Those codes are for *you* to type into *your* login screen. If someone wants you to speak it, they are trying to become you.

Credit freezes: the simplest way to stop ``new you" from being invented

Account takeover is terrifying because it's direct. New-account fraud is infuriating because it's bureaucratic. Someone opens a credit card, a loan, or even utilities in your name, and you discover it only when a bill collector calls you like you're the villain.

A credit freeze helps prevent that by restricting access to your credit file at the major credit bureaus. Without access, many lenders can't issue new credit in your name.

Freezing your credit is one of those rare life admin tasks that is both boring and profoundly effective. It doesn't stop someone from using an existing stolen card number, but it does make it much harder to open *new* accounts under your identity.

Two-income households have an extra reason to do this: your identity isn't just "yours." It's entangled with your partner's life. A fraud event can hit one person's credit and still disrupt the entire household—mortgage refinancing, car shopping, renting, even certain job background checks.

A good household move is to freeze credit for both partners and, if applicable, freeze credit for children as well (child identity theft is a real problem precisely because it can go unnoticed for years).

Unfreezing is usually straightforward when you actually need to apply for credit. The mild inconvenience is the point: it's a speed bump for criminals.

Your email is the master key

If you do only one "security upgrade," make it this: lock down the email account that controls your financial logins.

Because email is where password resets go. Email is where alerts go. Email is where the attacker goes after they get a foothold.

Treat your primary email like it's a vault:

- use a unique, randomly generated password stored in your password manager,
- enable the strongest MFA option available,
- review account recovery settings (make sure recovery email/phone numbers are yours),
- remove old devices and old app passwords if you see them.

If an attacker controls your email, they can quietly reset everything else while you sleep.

And in a two-income household, decide which email is the "primary identity" email. Many couples have a mishmash—one person uses an old college account for banking, the other uses a work email for everything, and alerts go to places nobody checks. Consolidate intentionally. You want financial alerts going somewhere that is monitored and protected.

Notifications: the early warning system that actually works

Most households don't need more "financial discipline." They need faster detection.

Turn on notifications for:

- large transactions,
- new payees or transfer destinations,
- password changes,
- login from a new device,
- withdrawals or wire transfers (if your bank offers it).

Yes, it's annoying. That's why it works. Fraud thrives in silence. Notifications add friction and visibility.

If you're worried about alert fatigue, route them strategically: one partner gets the account security alerts, the other gets the transaction alerts, or both get the high-severity ones. The point is that if something happens, someone notices quickly enough to act.

Household systems: how couples accidentally create risk

Digital defense isn't just about tools. It's about how two people share responsibility.

Here are the most common couple-level failure modes:

- *One person is "the money person."* Convenient, until that person is traveling, sick, stressed, or the target of a social engineering attack. Redundancy is safety.
- *Shared logins and shared MFA devices.* This creates single points of failure and makes it easier for a scammer to manipulate "the helpful spouse."

- *Unclear escalation.* When something looks suspicious, people hesitate. They don't want to bother their partner. They wait. Waiting is expensive.

A healthier model is "two keys to the vault." Not both partners micromanaging everything, but both capable of stepping in if needed.

One practical exercise: each partner should be able to answer, without rummaging:

- Where do our paychecks land?
- Where are our main investment accounts?
- How do we contact those institutions?
- Where are the passwords stored?
- What is the first action we take if we suspect fraud?

You don't want the first time you discuss these questions to be while staring at an unauthorized transfer.

If you suspect fraud: what to do in the first hour

In a true emergency, perfect is the enemy of fast. The first hour is about stopping the bleeding and preserving evidence.

A clean sequence:

1. *Stop interacting with the message/caller.* Don't argue. Don't explain. End contact.
2. *Use a known-good channel.* Call the number on the back of your card or from a bookmarked bank site. Not the number from the text or email.

3. *Freeze movement.* Ask to freeze accounts or disable transfers if needed. Change passwords from a clean device.
4. *Lock credit.* If identity theft is suspected, freeze credit immediately.
5. *Document.* Screenshots, dates, transaction IDs, the phone number used, any emails received.

Households often underestimate how much "containment" is emotional. Taking decisive steps quickly turns panic into a plan.

Actionable takeaway: build your household ``Digital Moat" in one afternoon

Pick a Saturday afternoon and do this as a household project, not a solo chore. Put on music. Make it slightly pleasant. Then build three layers:

1. *Credit freeze for both partners (and children if applicable).* Treat this as your perimeter fence against new-credit fraud.
2. *Password manager + unique passwords for your top 10 accounts.* Start with email, banking, brokerage, payroll/HR portal, and your mobile carrier account (because phone-number control matters).
3. *Upgrade MFA on financial and email accounts.* Prefer phishing-resistant options when available; otherwise use an authenticator app; leave SMS as a last resort.

Finish by writing a tiny "if-then" rule on paper and putting it where you'll see it:

> If we get an urgent message about money, we do not click. We log in through our saved link or call the number on the card.

That single habit—paired with strong login hygiene—turns most modern scams from a crisis into a harmless annoyance, which is exactly where they belong.